DATE			

Human Factors
Essentials

Human Factors Essentials

An Ergonomics Guide for
Designers, Engineers, Scientists, and Managers

Peggy Tillman
Barry Tillman

McGraw-Hill, Inc.
New York   St. Louis   San Francisco   Auckland   Bogotá
Caracas   Hamburg   Lisbon   London   Madrid
Mexico   Milan   Montreal   New Delhi   Paris
San Juan   São Paulo   Singapore
Sydney   Tokyo   Toronto

Library of Congress Cataloging-in-Publication Data

Tillman, Peggy
    Human factors essentials / Peggy Tillman, Barry Tillman.
        p.   cm.
    Includes bibliographical references (p.) and index.
    ISBN 0-07-064608-2
    1. Human engineering.   I. Tillman, Barry.   II. Title.
    TA166.T54   991
    620.8—dc20   90-23430
                                                        CIP

1 2 3 4 5 6 7 8 9 0   DOC/DOC   9 6 5 4 3 2 1 0

ISBN   0-07-064608-2

The sponsoring editor for this book was Harold W. Crawford, the editing supervisor was Caroline Levine, and the production supervisor was Suzanne W. Babeuf. It was set in Century Schoolbook by McGraw-Hill's Professional Publishing composition unit.

Printed and bound by R. R. Donnelley & Sons Company.

Dedicated to all the managers, designers, and engineers, who asked the right questions and made this book possible.

Contents

Part III   References

Preface

We wrote this book to help improve designs. Human factors is a profession that helps assure that equipment and systems are safe and easy to operate by humans. We want designers and product producers to use human factors. We want to expand the knowledge of human factors and demonstrate how to use it to improve designs.

As we count down to the year 2000, we approach a greater awareness of human needs. Consumers and product users demand that designers use human factors because consumers want products and their environment to be safe and comfortable. Designers and producers must meet these demands to avoid loss of sales, law suits, and catastrophic accidents.

Most everyone agrees, concern for the well being of humans is important, yet, often other design aspects come first. This happens for several reasons. Many designers and producers are unaware that valid, testable information and techniques exist for integrating humans into the design. Other designers and producers have been unsuccessful in attempts to apply human factors to a design. They may have hired the wrong professional or were unable to locate specific human factors information.

In this book we answer those questions that managers, designers, engineers, and other scientists have asked us about human factors. These answers explain what human factors is, how it can help your product, how to hire a human factors professional, and how to find and apply human factors information to specific design problems.

How to Use This Book

Part I: Answers to the Five Most Asked Questions

Part I will familiarize you with human factors and prove it can make your design safer and more effective. Chapter 1 clarifies what human factors is and why it is important. Chapter 2 demonstrates, with specific examples, why you should include human factors in your design process. The chapter explains why some designs work and some don't. In Chapter 3 we give human factors information sources and explain how they are obtained and validated. Chapter 4 describes when and how to hire a human factors professional. We explain the training for human factors professionals and how you can select the best professional for your project. Finally, if you hire a human factors professional, Chapter 5 will give you skills to manage a human factors project by supplying a checklist and detailed information.

Part II: Specific Answers to Help Improve Product Designs

Part II is for managers and designers who want to get directly involved in a human factors program. It supplies answers for specific human factors applications. You can use the chapters in Part II to manage a human factors professional or use the information to improve designs yourself. The chapters give an overview of the processes, tools, standards, and guidelines human factors professionals use. These chapters reveal techniques for applying human factors to designs. The last chapter gives you the facts you need to find specific information to improve your design.

Part III: References

Part III has information to help you find specific facts about human factors. It contains handy references for standards, human factors resource books, on-line information, and government agencies.

Peggy Tillman
Barry Tillman

Human Factors
Essentials

Answers
to the Five
Most Asked
Questions

1

What Is
Human Factors/
Ergonomics/
Human Engineering?

1.1 Definitions

Human factors is the process of designing for human use.

Consumers are demanding safe and effective products. However, not all people have control over products they use. Therefore, all products must be carefully designed. For example, if a child car safety seat fails because it does not fit the child or is difficult to install, everyone will lose: the child, the parent, the designer, and the manufacturer. Human factors can prevent these design disasters.

Human factors is a profession to help ensure that equipment and systems are safe and easy to operate by humans. A human factors researcher gathers and analyzes data on human beings (how they work, their size, their capabilities and limitations). A human factors engineer works with designers as a team member to incorporate data into designs to make sure people can operate and maintain the product or system. Human factors professionals then determine the skills needed to operate or maintain a finished product.

In recent years we have heard about accidents like Three Mile Island, Chernobyl, and the Belgian ferry sinking. Diligent human factors design may have avoided these accidents. As systems become more complex, designers must be careful to consider the abilities of individual users and operators. The human factors profession can help designers by supplying them with scientific information on human capabilities.

Humans have limitations and these limits must be respected when designing for humans and their environments. The human factors professional has analytical tools and techniques to help incorporate information on human capabilities and limitations into designs. Figures 1.1, 1.2, and 1.3 are from standardized government documents used by human factors professionals.

Human factors is difficult to define because it is a compilation of many sciences dealing with both humans and machines. For example, Figs. 1.1 through 1.3 illustrate research results from several scientific disciplines. Some of the disciplines human factors experts are trained in include the following:

Psychology   Anthropology

Engineering   Biology

Medicine   Education

Physiology

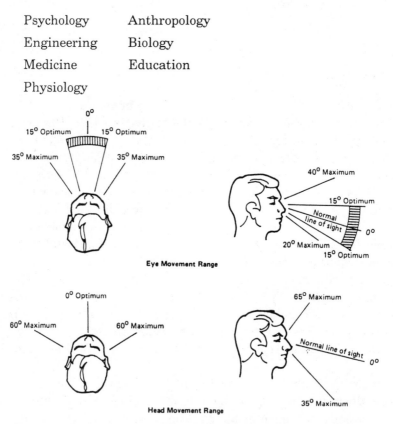

Figure 1.1   An example of human factors data. (*Reproduced from NASA-STD-3000, "Man-System Integration Standards," March 1987.*)

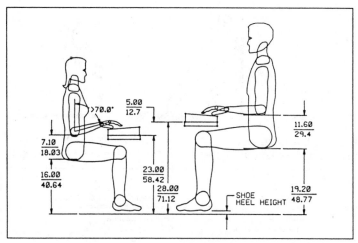

Figure 1.2   An example of human factors data. (*Reproduced fron ANSI/ HFS 100-1988, "American National Standard for Human Factors Engineering Visual Display Terminal Workstations," February 1989. Used by permission.*)

A human factors engineer focuses on the design of the equipment and systems while a human factors psychologist focuses on the human. Engineers define their job responsibilities differently than do psychologists. Because of this confusion about job responsibilities human factors professionals have not agreed on one name. Job titles currently in use include:

Human Factors

Human Engineering

Man-Machine Engineering

Ergonomics

Man-Machine Integration

Man-Systems Engineering

Human Factors Engineering

Biotechnology

Bioengineering

Engineering Psychology

Engineering Anthropology

TABLE XXIV. HORIZONTAL PUSH AND PULL FORCES EXERTABLE
INTERMITTENTLY OR FOR SHORT PERIODS OF TIME (MALE PERSONNEL)

HORIZONTAL FORCE*; AT LEAST	APPLIED WITH	CONDITION (μ: COEFFICIENT OF FRICTION)
100N (25 lb) push or pull	both hands or one shoulder or the back	with low traction $0.2 < \mu < 0.3$
200N (45 lb) push or pull	both hands or one shoulder or the back	with medium traction $\mu \sim 0.6$
250N (55 lb) push	one hand	if braced against a vertical wall 510–1525 mm (20–60 in.) from and parallel to the push panel
300N (70 lb) push or pull	both hands or one shoulder or the back	with high traction $\mu > 0.9$
500N (110 lb) push or pull	both hands or one shoulder or the back	if braced against a vertical wall 510–1780 mm (20–70 in.) from and parallel to the panel or if anchoring the feet on a perfectly nonslip ground (like a footrest)
750N (165 lb) push	the back	if braced against a vertical wall (600–1100 mm) (23–43 in.) from and parallel to the push panel or if anchoring the feet on a perfectly nonslip ground (like a footrest)

*May be doubled for two and tripled for three operators pushing simultaneously.
For the fourth and each additional operator, not more than 75% of their push
capability should be added.

Figure 1.3   An example of human factors data. (*Reproduced from MIL-STD-1472C,
"Human Engineering Design Criteria for Military Systems, Equipment and Facilities," Department of Defense, May 1984.*)

If human factors professionals agreed on one name, there could
still be confusion about the exact definition of that name. As Dr.
Roger W. Pease, Jr., Senior Science Editor at Merriam-Webster,
Inc., points out:

The definitions of "ergonomics," "biotechnology," and "human engineering" have been in need of revision because of changes in usage.

"Biotechnology" and "human engineering" have been used at times in the past as synonyms for "ergonomics." However, "biotechnology" has a new sense relating to molecular biology, which is backed by considerable evidence over the last ten years. "Human engineering" was once used to denote the activities of management spies, and in some quarters has or at least once had pejorative connotations. This may have been a contributing factor to its fall into relative disuse. We have defined it objectively in its first sense in the Collegiate—"management of human beings and affairs esp. in industry"—but in flipping through the citational evidence the historical aura of unpleasant associations is unmistakable.

"Human factors" has just not caught on yet in the general language, although it may be preferred in technical publications. Even in the U.S. and Canada the word "ergonomics" turns up with sufficient frequency to warrant the definition rather than a cross-reference.

Based on Dr. Pease's recommendations, the definitions of "ergonomics," "biotechnology," and "human engineering" were entered in revised form in *Webster's Ninth New Collegiate Dictionary* published in 1983 and appear as follows in the 1989 copyright[1]:

er·go·nom·ics \ˌər-gə-'näm-iks\ *n pl but sing or pl in constr* [*erg-* + *-nomics* (as in *bionomics*)] (1949) : an applied science concerned with the characteristics of people that need to be considered in designing and arranging things that they use in order that people and things will interact most effectively and safely — called also *human engineering* — **er·go·nom·ic** \-ik\ *adj* — **er·go·nom·i·cal·ly** \-i-k(ə-)lē\ *adv* — **er·gon·o·mist** \(ˌ)ər-'gän-ə-məst\ *n*

bio·tech·nol·o·gy \ˌbī-ō-tek-'näl-ə-jē\ *n* (1941)   1 : applied biological science (as bioengineering or recombinant DNA technology)   2 : ERGONOMICS — **bio·tech·no·log·i·cal** \-ˌtek-nə-'läj-i-kəl\ *adj*

human engineering *n* (1921)   1 : management of human beings and affairs esp. in industry   2 : ERGONOMICS

Dr. Alphonse Chapanis, past president of the Human Factors Society, International Ergonomics Association, and The Society of En-

[1]By permission. From *Webster's Ninth New Collegiate Dictionary* © 1989 by Merriam-Webster, Inc., publisher of the Merriam-Webster® dictionaries.

gineering Psychologists, gave the following definition for human factors.

Definition of Human Factors

Human factors is a body of knowledge about human abilities, human limitations and other human characteristics that are relevant to design.[2]

After 25 years of experience in human factors we have found no agreement on definitions. However, we believe there are two major human factors areas: research and application. The diagram in Fig. 1.4 shows how both research and application contribute to the human factors field.

The following definitions help clarify the distinctions between human factors research and human factors application.

Human Factors Research. Human factors research is a science of gathering information about humans. Human factors research professionals' goals are to understand human characteristics and how these relate to a product or system design. Human factors research also investigates human variables and makes information available to improve our knowledge of humans.

Human Factors Application. Human factors application is applying human factors data and principles to the specification, design, evaluation, operation, and maintenance of systems. Human factors application goals are to make products and systems safe, effective, and satisfying to use. "Human engineering," "human factors engineering," and "applied ergonomics" are additional terms used to describe the application of human factors information.

As Fig. 1.4 illustrates, human factors is a combination of two fields, research and application. The figure's inner circle (human factors) contains experienced human factors experts, tools, resources, data, and principles to improve designs. The research professions contribute information and data to human factors and the application professions use these resources to apply human factors to the designs.

For example, a nutritional research physician contributed to hu-

---

[2]Keynote address given by Dr. Alphonse Chapanis to the HFAC/ACE conference in Edmonton, Canada, on September 14, 1988.

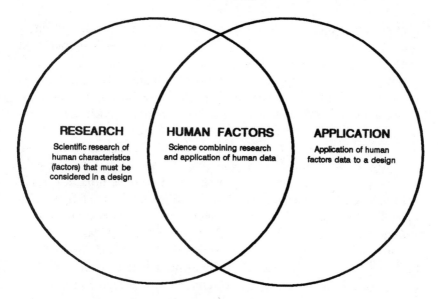

Figure 1.4   Two major areas of human factors.

man factors research by supplying NASA with nutritional requirements in a NASA human factors standard (NASA-STD-3000). An aerospace engineer then may apply nutritional information to designing a space station galley. The engineer can use the NASA human factors standard to assure long term space dwellers with an area to store, select, prepare, and eat nutritional meals.

In the following chapters we will use the above definitions. Chapter 4 provides descriptions and qualifications for the human factors professional.

1.2   A Short History

Human factors began as soon as humans designed equipment. Our ancestors selected rocks and clubs to fit their hands. At first, design flaws could be easily corrected or "lived with." For instance, if an ax handle was not comfortable, the user could reshape it, cut it shorter, wrap a cloth around the handle, or simply make a new handle.

Systems became complex. The need to consider human abilities became apparent with airplanes. Humans had to lower flaps, adjust throttles, and lower landing gear while concentrating on finding and aligning the airplane with the end of the runway. Confusion between two knobs could be disastrous. The military asked early human factors professionals to develop control knob shapes that could be identified by touch alone.

Aerospace human factors has grown tremendously. Humans are required to make life-threatening decisions in shorter time periods. Human factors expanded into complex systems, such as control rooms for nuclear power facilities, where human error can be catastrophic to thousands. Human factors professionals have helped design and locate controls and displays, write software, design jobs, and establish operator training programs to reduce human error.

In recent years, electronic technology filled homes with new products. The designers often did not consider the abilities of the consumer. Any modern home is likely to have a poorly designed television, video cassette recorder, compact disc player, microwave, burglar alarm, or computer that the owners do not understand how to operate. The human factors professional now stands at the beginning of a new area of responsibility: consumer products.

1.3   Human Factors Applications

Nearly all products and systems can be improved with human factors. A human factors professional may act as an expert witness in court and defend consumers rights to a safely designed product. Or, they may work for a large government contractor. The military requires the use of human factors experts on all large defense contracts.

The following are additional examples of ways human factors experts can improve products or systems.

Automobile design.   Human factors can improve automobile designs by making the automobile fit all sizes and shapes of people. A car interior must accommodate tall, short, overweight, very young, pregnant, and handicapped people. Effective human factors design can help lessen driver fatigue, improve visibility, and decrease accidents.

Schools.   Schools are improved by using human factors principles in planning and designing individual classrooms, restrooms, offices, sports facilities, meeting facilities, storage, and other special purpose areas. A human factors professional must consider the ages, physical sizes, and capabilities of individual users and how the facilities will be used in both the present and the future. Schools must accommodate all new technologies and social changes.

Business offices. Human factors helps business by designing and selecting equipment to make employees more comfortable and therefore more productive. They also help in designing and evaluating the office environment, such as lighting, heating, ventilation, noise control, and office space planning.

Personnel selection and training. The United States Army employs human factors engineers to design programs to select crewmembers for combat missions in helicopters. The human factors engineer is often asked to write training manuals and contribute information on the qualifications needed to operate equipment.

Computers. The human factors professional can help design computers and software for programmers or people using a computer for the first time. For example, a store clerk with little computer training may need to use a computer to ring up purchases, inventory stock, and supply a customer with a printout (receipt).

Manufacturing assembly plants. Robots are becoming common in assembly plants. Human factors professionals help by considering how the human works with automated machines that can cause physical harm to operators and maintainers.

1.4   Human Factors Methods

The human factors professional participates with engineers and designers in all design stages. Figure 1.5 shows system design stages and what human factors professionals do at each stage.

The first stage is predesign research. During this phase of the design, engineers and scientists research and experiment with ways to improve their designs. For example, a human factors professional might experiment with different controls for operating a robot.

The second stage is the preliminary design stage. Engineers produce and evaluate design ideas at this point in the design. The human factors professional participates in the preliminary design by determining the possible roles the human will play in the design: the crew size, the crew duties, and the skills required by the crew.

The third stage is the detail design and development. At this

DESIGN STAGE	HUMAN FACTORS ACTIVITIES	EXAMPLES
PREDESIGN ANALYSIS	Research and Test Human Characteristics (human factors analysts use the results in later design stages)	Active human factors research areas include: 1. Unique Control Systems (including voice activated controls and remote workstations with simulated operating environment) 2. Software Design - What is the best way for humans to work with computers? 3. Workload Measurement Tools - What factors contribute to fatigue and stress in nonphysical work activities (such as an office) and how can designers and planners avoid these factors?
PRELIMINARY DESIGN	Analyze Functions and Determine the Number of People and Skills Required to Operate and Maintain a System	NASA is planning a manned mission to Mars. How many people should go? The human factors professional can analyze the functions and estimate the number of humans required to operate the space craft, land on Mars, establish a base, explore the surface, and return home.
	Study Human and Machine Capabilities and Determine Which Tasks Should Be Done Manually and Which to Automate	A proposed system may develop an emergency that requires a response faster than human reaction time. The human factors analyst may choose a computer to respond initially to the emergency, make a temporary repair, and then alert the human to the problem.
	Define the Skills and Training Required by the System User and Establish Design Guidelines to Keep System Operating Complexity Within User Abilities	Computer hardware and software developers sometimes assume all users have comprehensive comput─ i s and knowledge. Fc· ─··· l ·, a computer company may want to develop software to manage a large farm. Human factors analysts can survey the education and computer skills of farmers and use the results as a guideline for software design and evaluation.

Figure 1.5   Human factors activities in systems design.

DESIGN STAGE	HUMAN FACTORS ACTIVITIES	EXAMPLES
DETAIL DESIGN AND DEVELOPMENT	Define the Environment Required for Operator Safety, Reduced Operator Fatigue, and Increased Performance	The human factors analyst defines the acceptable ranges for the environment including ventilation, noise, heat and cooling, lighting, vibration, and radiation. For instance, Human Factors professionals use tables to analyze a work facility, determine the jobs people do, and recommend lighting levels based on the job demands.
	Select and Design Controls and Displays	Controls and displays range from the computer keys and monitors in a missile launch control room to the controls to operate an elevator. For instance, designers must locate elevator buttons so they can be reached and read by people from three to seven feet tall, both sighted and blind. However, buttons may be accidentally bumped and the design should also account for this. The human factors analyst can arrange the buttons in a logical sequence and select and locate displays to show all passengers the floor selections and elevator position.
	Analyze Crew Skills and Task Demands and Prepare Operating Manuals and Training Materials	A soldier may have to operate sophisticated electronic military systems in a cold, dirty, dark, and stressful environment. Under these conditions, suitable training and instructions are very important. The human factors professional can prepare training materials and instructions to optimize the interface between the humans and machines.
TEST AND EVALUATION	Test the Human (Operators and Maintainers) in the Design	Human factors analysts use environmental testing, performance testing, questionnaires, and surveys to evaluate the ability of humans to operate and maintain a product. The results may uncover problems and suggest solutions. Many systems function poorly because the design does not account for operator abilities and limits. An unacceptable number of errors entered on computer keyboards, for instance, may be attributable to glare from the computer monitors. Observation and measurement of the work environment by human factors specialists can isolate these problems.

Figure 1.5   Human factors activities in systems design. (*Continued*)

stage designers select an idea and specify how the system will look and operate. The human factors professional participates in the process by making certain the humans in the design (the tank crew, for instance) can do their job effectively and safely.

In the last stage, designers test and evaluate their design. The human factors professional participates in the tests by focusing on evaluation of the working environment and performance of the crew.

2

How Can Human Factors Improve a Design?

2.1 Human Factors Professionals Are Champions for People

A few years ago I rented a car on a business trip. This car was so poorly human engineered that I needed to keep the operator's manual open on the seat for an entire week.

The windshield wiper control design became almost impossible to use and could cause a dead battery. The *OFF* position on the wiper control was in the center position and the different wiper setting speeds were on both sides of *OFF*. To make matters worse, I had to look behind the wheel to see the markings on the control. I could not feel the correct setting because there were no detents on the control wand. I actually had to pull off the road to find the correct setting for the wipers. In addition, the wipers worked with the ignition turned off and were in a location where they could be activated accidentally while the driver got out of the car. I discovered this problem the first night in the garage parking lot. I accidentally bumped the control and activated the intermittent setting. The wiper did not start while I was close enough to the car to notice. Fortunately I did return to the car and discovered the wipers on before the battery ran down.

Unfortunately the designer of the rental car excluded the human as part of the system. Mechanically the car may be very sound and efficient, but the human is part of the entire system. The designer must acknowledge the human's place in the design to make it an effective product.

Humans are a major component of systems. They are as important as electrical and hydraulic parts to the success of a system. Often, product producers will hire engineering specialists to design electrical or hydraulic components, but will not hire a representative trained in the needs of the human component. The human needs a champion, particularly when a system or product is complex or has dangerous operations involving humans. Human factors professionals also help the design team by winning product user acceptance. They accomplish this by examining what the user needs to operate and maintain the product or system.

This chapter shows ways human factors helps designs. The final section discusses actual products and the effect of human factors on the design and use of these products.

2.2   Function and Tasks Analysis

The human factors professional uses scientific analytical approaches to improve designs. One analysis method we use is a design standard from the Department of Defense, MIL-H-46855, "Human Engineering Requirements for Military Systems, Equipment, and Facilities." According to this standard, human factors analysis has three basic steps.

The first step in this analysis is to define and describe the system's objectives or functions. Examples of system functions are launching a missile, exploring the moon surface, or mining coal.

The second step in the human factors process is to determine the best way to accomplish the system's functions or objectives. The human factors analyst helps determine which activities should be done by humans and what jobs should be done by machines. Often humans are more flexible and more efficient than a machine. For instance, a human inspector is often the best way to check product quality before shipping the merchandise to the customer. Other jobs may be too boring, too dangerous, or too physically demanding for a human. For example, we use robots to explore other planets in our solar system. Human factors analysts know human capabilities and limits and assign functions to either the machine or human to best achieve the system or product objectives.

After assigning duties and responsibilities to humans in the system, the third step is to examine human tasks or jobs in detail. Task analysis examines human perceptual, judgmental, motor, and emotional demands for each job or task operation. The human factors professional can use information obtained from functional and task analysis to determine:

- The most efficient task step sequence
- People and skills required to do the tasks
- Potential safety problems and ways to minimize the hazards
- The most effective controls, displays, communication systems, and workstation layout
- Environmental requirements such as lighting, noise, heating, and ventilation
- Requirements for labels, operating and maintenance manuals, and training

The following example illustrates how human factors function and task analysis contribute to a design. The space shuttle uses a robot arm remotely controlled by astronauts inside the shuttle to deploy satellites. In the hostile and airless environment of space some jobs should be done by robots instead of astronauts. Repairs and construction conducted outside a space capsule is slow, awkward, and dangerous for humans. Besides safety factors, there are cost considerations. NASA estimates that it cost $2500 per crewmember minute in space. The cost of putting on and removing space suits is high. As a result, NASA chose to allocate some jobs to remotely controlled robots. The human factors analyst helped decide when and how to use humans and gave jobs that exceed human capabilities to machines.

After the design team has decided which jobs to allocate to equipment and which ones to humans, human factors professionals analyze tasks and help design the system to enhance human performance. In our example, NASA allocated the functions of launching a satellite to a remotely controlled arm. The human factors analyst then determined what the human must do to operate the arm. These operations include physical actions, visibility requirements, and mental skills. Human factors specialists work with designers to select controls to operate the robot arm, place the windows for the operator, and design video cameras and monitors necessary for precise operations.

Part II explains how to perform functional and task analysis in more detail.

2.3   Provide Information to Designers

Both function allocations and task analyses define what humans must do to operate and maintain a system. This information shows the human factors professional where to apply human factors design standards and guidelines. The human factors professional can begin by working with the project team to solve detail design problems. For example, functional and task analysis showed that our astronaut had to handle objects 50 ft away with the robot arm. Human factors analysts determined that the astronaut needed video cameras to see the arm operation in detail. The human factors analysts then provided designers information including the required display resolution, monitor and camera placement, monitor size, and lighting requirements.

The human factors profession obtains data from research, standards, and guidelines. Human factors standards and design handbooks are a compilation of information that specialists have successfully applied to systems. For example, all military equipment must meet the human factors standards in Military Standard, MIL-STD-1472, "Human Engineering Design Criteria for Military Systems, Equipment, and Facilities." The following are general requirements in this standard:

Objectives

Standardization

Function allocation

Human engineering design

Fail safe design

Simplicity of design

Interaction

Safety

Ruggedness

Design for NBC survivability

A few of the detailed requirements in this standard are:

Cathode ray tube (CRT) displays

Verbal warning signals

Prevention of accidental actuation

Touch screen controls for displays

Anthropometric data

Common working positions

Test equipment

Electric, mechanical, fluid, toxic, and radiation hazards

The human factors information in the standards can be used directly by designers. Sometimes, however, the human factors professional needs to select and interpret appropriate data. More information on design standards and handbooks is in Part II of this book.

2.4   Testing

The human factors professional tests, observes, and records human performance. Human factors testing determines where human problems occur in a design. Testing can be as simple as taking a noise level reading, or as complex as evaluating mental workload on an air traffic controller at a busy airport. During predesign a human factors professional works with the designer to test design ideas, to note if people perform as expected, and to determine possible error sources.

Human factors professionals validate if people can operate and maintain the design by testing the finished product or system. Human factors testing can isolate problems in an existing design and help determine ways to reduce human errors.

The following examples show how testing improves products or systems.

The Soviet space program recently launched a solar system exploration probe. The human controller on Earth made an error in sending commands to the probe. The controller mistakenly signaled the probe to turn its antenna away from the Earth. The probe responded to the signal and turned away from the Earth, not allowing reception of any further signals from the Earth. By testing human performance, designers may have identified potential operator errors and prevented the probe loss.

The military conducts missions in arctic conditions. Arctic clothing changes the shape, size and range of motion of a human. We as human factors professionals have put on arctic clothing in 100°F weather and have tried tasks with the restrictions of the arctic clothing. It is almost impossible to grab a normal-sized metal handle while wearing an arctic mitten. Grabbing a metal handle without a mitten in subzero weather can cause a painful injury. Arctic clothes add roughly 2 in to body width. As Fig. 2.1 shows, a vehicle hatch that allows a large man to fit through in normal clothing may not allow a smaller man in arctic clothing.

We found that voice communications in military vehicles, even with electrically aided headsets, is very poor because of the extremely high noise levels. The noisy environment prevents the commander from relaying complex information to the troops. We used instruments to test the communications system and calculate the articulation index. One solution to communication difficulties was to limit the vocabulary so troops only need to recognize a few hundred words. This involved no hardware redesign.

Extra vehicular activities (EVA) outside a space vehicle is very expensive because of complex space suits and support equipment. NASA does extensive simulation of astronaut activities before the mission to streamline tasks and make the astronaut as efficient as possible. NASA uses water tanks to simulate actions in zero gravity. The test results help in the design of equipment and in the scheduling of human tasks. The task scheduling is important because humans need life support systems (oxygen, cooling, warmth, radiation protection) in space. Because some tasks cannot always be stopped

STATURE OF MALE (INCHES)	APPROXIMATE SHOULDER WIDTH (INCHES)	
	WITH ARCTIC CLOTHING	WITH SUMMER SHIRT
67	(20)	18
72	22	(20)

Figure 2.1   Ingress and egress clearances for males wearing arctic clothing.

when the human runs out of life support, it is important to run tests or simulations to determine how long a task takes to complete.

2.5 Prepare Operating and Maintenance Manuals

A goal of the human factors professional is to make operating and maintenance instructions clear and complete. Human factors standards exist for layout, organization, and content of labels and written instructions.

The human factors analysis provides product designers with information to define procedures for operating or maintaining systems or products. The information includes the most efficient sequence of activities, the type of user, the appropriate reading grade level, and environmental conditions. For example, high stress emergency conditions or poor lighting will influence the content and form of the instructions. Information obtained from the analysis process can be used to write instruction manuals, labels, placards, and training programs.

2.6 Safety

Product litigation is becoming commonplace. In recent years attorneys have used human factors professionals as expert witnesses, often proving that an accident was caused by poor design. The Forensics Professional Technical Group of the Human Factors Society had 283 members as of December 31, 1988 (see Chapter 12). By making the human an important consideration in a design the manufacturer can prevent costly litigation and possible redesign.

Because humans cause accidents and are injured by accidents, the human factors profession is concerned with safety. Below is a list of four commonly accepted ways to reduce hazards and how human factors can help with each. The hazard reduction list below is formally documented in the Department of Defense safety standard MIL-STD-882, "System Safety Program Requirements."

1. *Eliminate the hazard through design:* Human errors cause accidents. Human factors can help design controls and displays that are simple to understand and operate. Human factors can plan activities in a system to reduce errors caused by overwork or boredom.

2. *Incorporate safety devices:*   Human factors can help with design and location of safety devices such as emergency shutoff controls. Human factors can also provide dimensions for the proper fit of safety goggles, gloves, and respirators.

3. *Provide warning devices:*   Human factors professionals can help determine the color, location, and wording of warning devices, the volume and pitch of warning signals, and the design of warning and caution markings on gauges and video displays.

4. *Develop procedures and training:*   When hazards cannot be eliminated in other ways designers must rely on the human to follow proper procedures to avoid danger. The human factors professional knows human limits and can help make the procedures simple and easy to remember. Human factors can also establish criteria for personnel selection and development of safety training programs.

The following examples show how human factors can improve safety.

Jobs that undertax the capabilities of operators can cause the human to make errors. For example, processes are often automated in chemical plants and operators must monitor temperature gauges that almost never move. These operators will get bored and do other things: get coffee, play ping pong, sleep. Terrible accidents can and have occurred under these conditions. Human factors professionals have recommended automating boring jobs and allowing humans to do more productive jobs like setting up a batch of chemicals for a new mix. Most chemical plants now have automatic controls that stop a dangerous condition and activate warning tones and lights to alert the operator. Human factors specialists can define the frequency and amplitude of the warning tones in a noisy environment. Visual warning indicators should be bright and located to attract attention. Controls should be coded so the operator can quickly assess the situation and respond.

Many military vehicles have automatic fire sensors and extinguishers because the vehicles can overheat and catch on fire during heavy use. However when these extinguishers go off, they cause the engine to quit. This may be acceptable under normal conditions, but not in combat. The soldiers may choose to let a small fire burn and continue to use the engine to retreat. Human factors specialists helped design controls to change the extinguisher from automatic to manual operation. Human factors specialists worked with designers to design indicators that let the vehicle operator know about a vehicle fire and the extinguisher system status (automatic or manual).

Coal miners are constantly in danger of a roof collapse. The human factors profession helped design mining vehicles to allow the miners to continue productive mining operation with improved safety. The miners now use vehicles designed with protective cabs and controls within the miners' reach so that they do not have to leave the cab protection. The cabs are small enough to fit in low tunnels yet still have enough room for the largest miner to operate all the controls. The cabs also allow unrestricted escape in an emergency.

2.7   Product Redesign

Not every design works. Many times a product is unsuccessful because it does not do what the user expects. Often a product appears to fail because of a mechanical failure, when the failure really is human error. A human factors professional applies the same analytical techniques used on a new design to determine the product's problems and makes recommendations for improvement. The recommendations may be as simple as a warning label or clearer operating instructions.

The redesign and resolution of human factors problems can be made for vast complex systems or products as small as a kitchen egg beater. A major hotel chain built a 319 room downtown hotel with large convention facilities. Although the hotel was attractive in most respects, business was poor. After staying in the hotel it became apparent that there were many human factors problems that made the hotel uncomfortable. One problem was the location of their parking lot, a city block and a flight of stairs away from the check-in desk. The bell captain was stationed at the front door where there was only enough parking for hotel vans. If this hotel stationed hotel employees at the door to the parking lot this customer inconvenience could be eliminated.

Human factors can help redesign products to make them more marketable. In the 1970s both microwave ovens and new displays became available to consumers. Light emitting diodes and liquid crystal displays became readily available and designers went wild with the push buttons and newly available displays. In the name of technical innovation the designers of microwave ovens overwhelmed the consumer with complex controls and displays not needed to operate a microwave. This was only later recognized because microwave ovens did not sell the way the microwave industry had hoped. The consumers still do not use all the available functions or even do the wide range of cooking tasks possible because they just do not understand how to use them.

After producers marketed digital watches for several years we see the reintroduction of the analog watch. The jump in technology that allowed number displays excited designers. However, digital watch owners had to do mental computations to determine how many minutes before lunch. It is often simpler to glance at a watch hand to judge the distance to noon and interpret that distance into how much time is left before lunch. After using the new digital technology for many years, many consumers realized it was easier to judge time with hands on a dial.

The automobile speedometer and fuel gauge should not require drivers to make mental computations or learn complicated instructions to determine how fast they are going or how many gallons of gas they have. Information displays should not make people do computations while operating vehicles.

Poor human factors is apparent in many products. Humans will "make do" by making their own adjustments. Often, the user redesigns equipment. These are the types of design errors that the human factors professional tries to eliminate before final production. Examples of these user redesigns can be seen on machinery in a shop or on construction equipment. The operators have recognized that they have made errors and make modifications or warning notes on the equipment. Typical corrections or modifications might include padding of a sharp corner with a piece of tape or foam or pasting on a new set of instructions. The instructions might rename an item because the engineer labeled it by a term not commonly used by the operators of the equipment. A control panel light might be too bright or annoying and an operator may put a piece of tape over it. This light could be a warning light that was distracting and useless because it continued to stay on even though an operator corrected the malfunction. Extensions on handles to obtain more leverage is another example of user redesign. These are graphic illustrations of problems the human factors person will look for and correct.

While doing human factors redesign for tanks we found excellent information by going out to the vehicle test track and looking at the tanks that had been driven around for weeks. There we found padding on sharp corners and straps screwed on where a person could hang on and maintain balance while bouncing over rugged terrain.

It is important in the design and incorporation of new technology to remember how people see and understand machines.

2.8 Examples of Good and Bad Human Factors Designs

2.8.1 Automobile shoulder belts

Failure to include human factors information in a design can result in serious injury or even death to the user. Many auto shoulder belts do not adjust for the different adult and children heights. As a result, a shoulder harness may lie across the neck of a short adult and give no protection to children. The photos show a 6-ft male and 5-ft 5-in female (Fig. 2.2) in the driver's seat of the same auto. As you can see the male is able to drive with the belt in the proper position. The female (with a short torso) has little choice but to have the belt ride across her neck. She is uncomfortable and the belt could cause severe injury to her neck if she was thrown forward in a collision. The problem is even worse for children too big for toddler seats. It is not surprising many people simply refuse to wear a shoulder harnesses. Unfortunately, most lap belts and shoulder harnesses are interconnected and removing the shoulder belt leaves the person totally without crash restraint protection. Fewer than half the adults and approximately one-third of the children between ages 5 and 12 wear safety belts[1].

The second set of photos (Fig. 2.3) show the same drivers in a car with adjustable shoulder belts. The top of the shoulder belt is attached to a bracket that moves up or down allowing the driver or passenger of the car to adjust the belt with one hand. Unfortunately for short adults and children, few cars include this simple solution to a human factors safety problem.

The female model in these photos was recently in a minor accident and received bruises across her neck from the safety belt. A more severe accident may have crushed her trachea.

---

[1]U.S. Department of Transportation, *Restraint System Usage in the Traffic Population*, Rep. DOT HS 807 343, Washington, August 1988.

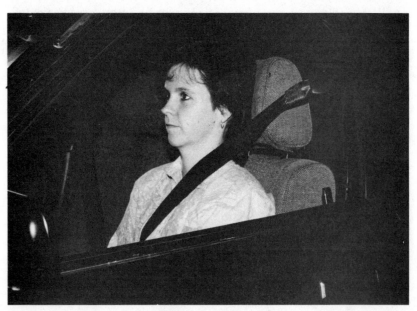

Figure 2.2   Nonadjustable shoulder harness cuts across the neck of a small person (bottom).

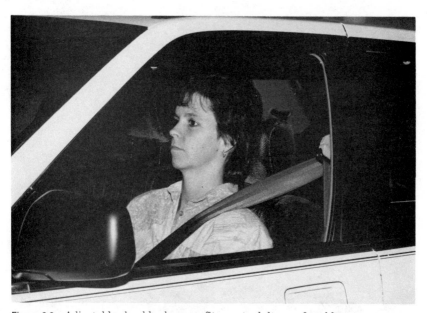

Figure 2.3   Adjustable shoulder harness fits most adults comfortably.

2.8.2 Computer handle

Figure 2.4 illustrates a human factors flaw in the design of a laptop computer. The computer functions excellently, but the handle design makes the computer difficult to pick up. The handle folds neatly out of the way on the bottom of the computer and is held there by friction. However, once you fold the handle back under the computer into its slot, the handle becomes difficult to pull back into the carrying position. There is an indent (arrow) in the case next to the folded handle but you cannot get your finger under the handle to pull it out of the slot. Designers could have prevented this by checking the human factors standards for finger clearance.

We found a solution by putting a ribbon around the handle. When we pull on the ribbon we pull the handle out of the slot and grab it. Our solution is an example of how consumers often are forced to make changes to a product. At times these changes are minor like our computer handle and don't effect the product's intended use. But often the user makes changes in a product that could harm the product or themselves.

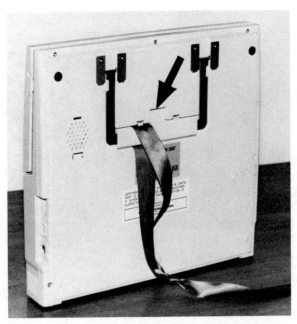

Figure 2.4   Poor laptop-computer handle design.

2.8.3 Nuclear power plants

Poor human factors designs affect everyone. This fact became apparent in 1979 when the nuclear power plant at Three Mile Island failed. The displays in the plant did not adequately explain what was happening and the plant came within an hour of a catastrophic meltdown. The National Academy of Sciences Research Council in a report released in February 6, 1987 stated that the NRC had neglected to examine the "human factors" that could lead to a nuclear accident.

Two years before the Three Mile Island failure, human factors professionals reported that designers of these complex and potentially dangerous systems frequently ignored human factors design principles. This report prepared for the Electric Power Research Institute[2] was largely ignored by the Nuclear Regulatory Agency. The photos in Figs. 2.5 to 2.7 show a few problems identified in the 1977 report.

Figure 2.5 shows a technician forced to use a chair as a writing surface. This lack of writing surface could cause operator fatigue, transcribing errors, and decreased motivation for making proper recordings. The second photo (Fig. 2.6) shows a control just a few inches from the floor. The control could be easily kicked or broken. The control status lights, which are in the same location, could go unnoticed because they are out of the operator's normal field of view. The last photo (Fig. 2.7) shows a technician changing a bulb in a status indicator. Standing on the sloped panel is dangerous but is the best option to reach the bulb. Leaving a burned-out bulb may eventually cause a disaster. Improperly human factored status lights were cited as a cause for the accident at Three Mile Island.

2.8.4 Printer

Consumers refuse to buy products that ignore human factors. Products easy to understand and use ("user friendly") sell better than products that ignore what the user understands and is willing to learn. We purchased a computer printer that performs all the functions anyone could ask for, but is very difficult to operate. You can select the printer options (font, pitch, etc.) from the front panel but

---

[2]EPRI NP-309 Project 501 Final Report, Human Factors Review of Nuclear Power Plant Control Room Design, Prepared by Lockheed Missiles Space Company, Inc., Sunnyvale, California, March 1977.

Figure 2.5   No writing surface forces a nuclear power plant technician to use a chair. (*Photograph used by permission of Joseph L. Seminara.*)

you must use a 60-page manual to learn the printer's operation. To solve this problem in our office we made a program list and step-by-step instructions on how to program the printer and placed our list under the printer (Fig. 2.8). Our personal instruction manual helps, but it only covers our routine use, and therefore, many other printer options go unused. Designers involved with complex problems behind the control panel often do not look at the control panel from the user's side. A good instruction manual can help, but does

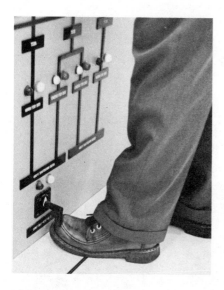

Figure 2.6 Low placement of a nuclear power plant control allows it to be kicked and broken. Also it is difficult to see lights and labels. (*Photograph used by permission of Joseph L. Seminara.*)

Figure 2.7 Nuclear power plant technician changing a status indicator bulb. (*Photograph from EPRI NP-309 Project 501 Final Report, "Human Factors Review of Nuclear Power Plant Control Rooms," Lockheed Missiles and Space Company, Sunnyvale, Calif., March 1977. Used by permission.*)

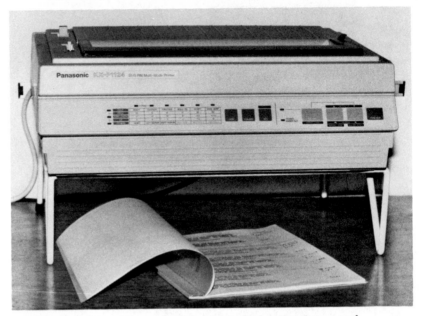

Figure 2.8   Computer printer with our supplemental instruction manual.

not substitute for an effective design. Someone must represent the interests of the human component of the system.

2.8.5   Car trunk

Human factors research shows the limits people can lift. Generally strength decreases the higher we lift something. Many automobile trunk openings are too high and make loading a heavy item such as a suitcase difficult. The photos in Figs. 2.9 and 2.10 show the difference between lifting a suitcase into a trunk that opens to the bumper and a trunk with the opening 1-ft higher. The higher trunk height can mean that the suitcase needs to be 10 lb lighter. A car with a lower trunk opening shows a designer's awareness of the capabilities of different sized users.

2.8.6   Infant car seat and carrier

We discovered an infant car seat and carrier that is an excellent human factors design example. This consumer product manufacturer has thought through the users' needs. The seat combines two functions: It is a convenient carrier for the child (Fig. 2.11) and can

Figure 2.9   A high-lip trunk opening makes loading difficult.

Figure 2.10 A low-lip trunk opening makes loading easier.

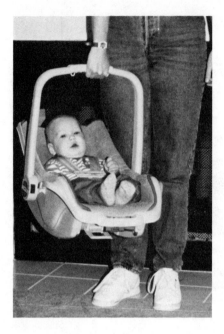

Figure 2.11 The infant carrier can be carried with one hand.

be easily buckled in a car to meet auto infant seat safety laws (Fig. 2.12). The parent or guardian does not have to switch the child from one seat to another.

Features include a large padded handle for carrying. When placed on a flat surface the seat rocks or can be made to sit in a

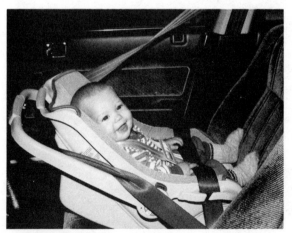

Figure 2.12 The infant car seat instructions are prominently displayed to help proper installation.

stable position for feeding. The handle folds back and locks to stabilize the seat. The seat comes with a tray and a canopy. At the back of the seat is a compartment for miscellaneous items including the seat instructions and the folded canopy. Under the seat is a clip for a bottle. There are two handles (or hand grips) at the ends of the seat to help position it in the car. This seat also meets FAA airplane standards.

A survey published in August 1988 by the National Highway Traffic Safety Administration (DOT HS 807 342) found nearly 30 percent of auto infant seats are incorrectly installed. As seen in Fig. 2.12, the car/carrier seat has illustrated instructions placed prominently on the side of the seat to help ensure proper installation in the automobile.

The seat is properly human factored because the designers thought through the tasks of transporting and caring for an infant. The seat meets the comfort and safety needs of the infant and the parents.

3

Where Does Human Factors Information Come From?

3.1 Information Paths

This chapter explains scientific information sources and paths human factors data takes to become incorporated into a design. Literature references and human factors information addresses are in Part III.

It is dangerous to believe because we are human, that human factors data is obvious and doesn't require research. A human factors professional does not measure his hand or foot to determine glove or shoe sizes (Fig. 3.1). They understand that one test subject alone is not a valid data source when designing for an entire population, even if the one person was "average size." A perfectly "average-sized" person does not exist. It is an error to design for this nonexistent person. If a clothing manufacturer tailored for only "average-sized" people their clothes would fit only a few customers. Yet often equipment designers design for the "average-sized" user. The human factors professional knows a design must fit a range of population sizes and human factors data helps designers accommodate all people.

The human factors researcher compiles data from many professional fields and information sources. These include performance tests, physical measurements of humans, task observations, and

Figure 3.1 This is NOT the source of human factors information.

surveys. After obtaining research results, human factors specialists either apply them directly to a design or use the results to develop guidelines and standards. The reach height of adults is an example of a human factors guideline. This information allows designers to put controls and handles within everyone's reach. Other guidelines include room temperature and humidity limits at which humans make excessive errors in tasks such as controlling air traffic or writing a research report.

Figure 3.2 illustrates the human factors research process and how the research results are applied to the design.

3.2   Information Sources

Research on humans is the source of human factors data. As Fig. 3.3 illustrates, research is one component of human factors. Human

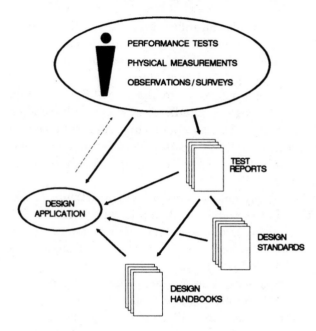

PERFORMANCE TESTS

PHYSICAL MEASUREMENTS

OBSERVATIONS / SURVEYS

TEST
REPORTS

DESIGN
APPLICATION

DESIGN
STANDARDS

DESIGN
HANDBOOKS

Figure 3.2   The flow of human factors information from research to application.

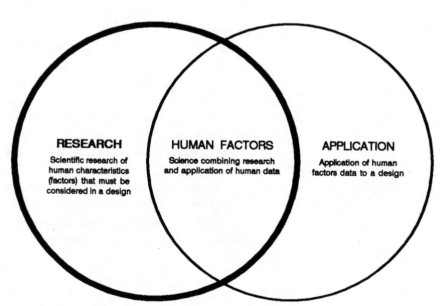

RESEARCH

Scientific research of
human characteristics
(factors) that must be
considered in a design

HUMAN FACTORS

Science combining research
and application of human data

APPLICATION

Application of human
factors data to a design

Figure 3.3   Research is the source of human factors data.

characteristics research builds the information foundation designers need. Humans are very complex systems and valuable human factors information evolves from varied sciences including medicine, anthropology, psychology, and physiology.

For studies on human characteristics to become valid information and aid in the design process, researchers must correctly conduct and document their studies. A researcher's education includes how to conduct tests and interpret test results. Because humans vary, researchers make certain the sample size is large enough to be statistically significant.

Universities, the government (including the Department of Defense, National Aeronautics and Space Administration, and the Department of Transportation), and industry fund and publish research. Often research findings are available to designers in professional journals. In addition to the formal research reports, researchers conduct experiments and studies to answer specific design questions. Researchers often report these studies informally and file them in company files instead of publishing the results in scientific journals.

Following are categories of human factors research that show how researchers conduct studies to provide useful information for systems design and development.

Human Performance Measurements

Researchers set up test situations to measure human mental or physical performance in different environments. The photographs show what a pilot sees from an airplane cockpit. During landings the pilots visual focus must change frequently from the instrument panel (close focus) to the runway (distance focus). Human factors researchers are investigating dangerous refocus times to help reduce human errors. The first photo (Fig. 3.4) shows what pilots see as they view the instruments. Notice that the runway is out of focus. The second photo (Fig. 3.5) shows the pilot's focus on the runway. The researcher found it took 2.5 s for the pilots to refocus from close focus to distance focus. Research of this type provides information to help designers select safer displays, controls, and procedures.

Physical Measurements

Human factors researchers measure the lengths and weight of people. The U.S. Army recently sponsored and conducted a survey

Figure 3.4   Example of human factors research on aircraft pilot's visual focus. Photograph shows pilot's vision when focused on instruments. Notice how the view outside the airplane is out of focus. (*Photograph used by permission of Dr. Richard Haines.*)

Figure 3.5   Example of human factors research on aircraft pilot's visual focus. Photograph shows pilot's vision when focused outside airplane. Notice how the instrument panel is out of focus. The time to shift from close focus (Fig. 3.4) to distant focus is roughly 2.5 s. (*Photograph used by permission of Dr. Richard Haines.*)

of the physical dimensions of our soldiers. This anthropometric information is needed to design equipment such as seats and hatches in tanks or clothing and personal gear such as helmets or gas masks. A team of 22 anthropometric specialists spent 1 year measuring 132 body dimensions on nearly 9000 subjects.

In designing facilities for an elderly population, such as a convalescent home or retirement community, an architect might want to know the maximum resistance to build into a self-closing door. If human strength data is not readily available, the architect may conduct a research study. Researchers should test the proper population (elderly) to determine the information. Strength varies with the type of action (pull, push, lift, etc.) and the direction of the force. Researchers must test the strength in the direction needed to open a door.

Observations and Surveys

Much of human factors research is done "in the field" by observing and interviewing people. We have interviewed, observed, and filmed workers to help design better workstations and controls. NASA researchers have spent months in Antarctica isolation to determine the confinement effects for long-term space missions.

In summary, human factors information comes from research in many scientific fields. Researchers are careful in designing studies to obtain valid results. Research results can become a foundation for further research, or results can be directly incorporated into a design. Some human factors information is transferred to human factors design handbooks and to human factors design standards. The following section explains how to use research reports, design handbooks, and human factors design standards.

3.3 Specific Sources of Human Factors Data

As Fig. 3.2 illustrates, test reports, human factors design handbooks, and human factors design standards contain the results on human characteristics. The following paragraphs describe each of these human factors information sources and how a systems developer can use them.

3.3.1 Research reports

Research reports contain both results and research techniques. Research techniques often involve sophisticated measurement technology and statistical tools. Reports give careful explanations of experiment techniques so that other researchers may reproduce the study if they need to confirm the validity of the findings or do further research.

Some research projects answer a specific design problem and the designer can apply the research results directly to a design. These results are usually reported informally because the researcher did not receive funds to prepare a formal report. Sometimes a design organization will formalize these studies and distribute an "in-house" document, but often only an informal report is sent to the designer who requested the information.

3.3.2 Human factors design handbooks

Design handbooks and guidelines condense human factors research results and studies that are applicable to designs. Authors of these handbooks will put research results into charts and tables for designers. Designers have used and validated these guidelines.

Design guidelines are constantly updated and expanded. Human research develops and incorporates new data into design guidelines. The U.S. Army anthropometric survey is an example. The size of the Army population is changing and current survey results will displace earlier information. Ten years ago human factors design standards for computer hardware and software did not exist. With the rapid change in technology, design handbooks must now contain useful research conducted on human-computer interface.

3.3.3 Human factors design standards

Human factors design requirements are applicable to a wide variety of situations and are essential to safe and effective human performance. Human factors design standards exist for many government and commercial products. Some organizations that have developed or use human factors design standards are the Department of Defense, NASA, and the Society of Automotive Engineers. These design standards are based on human factors research and testing. Most design standards are divided into two parts:

Recommendations: Recommendations are information that should be considered in a design but is not necessary to follow exactly. The word used in a standard to indicate recommendations is "should."

Requirements: These are human factors design standards that must be followed. Research and experience has confirmed that these requirements are necessary for systems to be safe and productive. The word used in a standard to indicate a requirement is "shall."

Design standards are simple to follow and usually require very little interpretation by a designer. This is the first human factors requirement a designer incorporates in a design. Design standards do not always cover all human factors problems. For this reason the designer and human factor professional also must rely on test results, design guidelines, and handbooks besides design standards.

4

Do I Need
a Human Factors
Professional?

4.1 Human Factors Knowledge
Improves Designs

We as human factors professionals would like you to believe the answer to this chapter's question, "Do I need a human factors professional?" is a resounding "Yes!" In reality, improvement in a design can be made by anyone using available human factors knowledge.

We are sure human factors professionals would rather have designers use human factors knowledge available to improve a design than not consider the human at all. If you or your company believe it is not feasible to use a human factors professional, Part II and Part III of this book will help you to incorporate the human into your design.

Whether your project needs a human factors professional depends on the project and the contractual obligations. Items to look at in judging the project needs are:

Contractual requirements: Many military and other government contracts require a human factors professional to work full or part time on a project.

Past human factors problems: Workstations in a large telephone service company were responsible for operator fatigue and physical ailments. It was necessary to involve human factors professionals to resolve the problems.

Special problems: Display and control systems in high performance military aircraft often require human factors specialization.

Reliance on human performance: Some systems depend greatly on the human performance for success. For example, automobiles can kill if drivers do not do their job correctly.

An alternative to using a human factors professional full time, is to hire one for specific project phases. Ideally, a human factors professional should help with the design from the concept phase through the redesign of the product. Many human factors professionals are hired on a subcontract basis to solve specific problems at any phase of the design. Human factors professionals are available for short-term contracts. This person or company should be willing to sign a contract or statement of work including a concrete deliverable or produce results for a specific amount of money and time.

4.2   What Kinds of Human Factors Professionals Are Available?

If you decide to hire a human factors professional you must hire the right person for the project. This person must be able to supply input to the design and improve the final product. This input must be realistic and measurable and based on solid research.

4.2.1   What is a human factors engineer?

In hiring a human factors person you will need to understand what qualifies people to call themselves human factors professionals. These professionals use terms like human factors engineer, human factors analyst, human engineer, and ergonomist. There are important distinctions among these titles that you need to understand to obtain the best human factors professional for your job or project. The person's educational background is one way to clarify these distinctions.

Until recently there were no degrees given for human factors and very limited college courses were available. People interested in the field received degrees in varied disciplines, the most common being psychology. The United States Human Factors Society 1990 membership illustrates this with 44.6 percent of members holding degrees in psychology (Fig. 4.1).

COLLEGE DEGREE	PERCENTAGE OF HUMAN FACTORS SOCIETY MEMBERS
Psychology	44.6 %
Engineering	17.5 %
Human Factors/Ergonomics	7 %
Industrial Design	2.8 %
Medicine/Physiology/ Life Sciences	4.1 %
Education	2 %
Business Administration	2.9 %
Computer Science	1.3 %
Other	7.1 %

Figure 4.1   College degrees of the United States Human Factors Society members in 1990. (*From "The Human Factors Society Directory and Yearbook," 1990.*)

Educational diversity makes it difficult to hire the correct human factors professional for the job. By selecting the correct educational background, your chances of hiring the person that will improve your product the most will be improved. You must carefully check the individual's professional training to determine exactly what the person is capable of doing.

A human factors analyst is usually an individual with a degree in human factors, psychology, or other disciplines. The human factors engineer or human engineer, as the name implies, has a degree in one of the engineering fields. Many human factors professionals without engineering degrees identify themselves as human factors engineers. If people call themselves human factors engineers but have no engineering background, a problem could develop if you expect them to work with engineers.

4.2.2 Research versus applied

Your project will dictate what education and experience the human factors professional must have. If your needs are to obtain research then your best choice is a human factors analyst or an individual with research or psychology degrees. If you need to apply human factors information to a design, then you should hire a human factors engineer or one with an engineering degree.

We witnessed one example of a person with the wrong educational background working for a contractor who produces military hardware. The contractor hired a person whose title was "Human Factors Engineer," but whose educational background and experience was in research psychology. This professional began designing tests, rounding up test subjects, gathering detailed data, and interpreting the results.

The project was in the final design stages and the product designers needed immediate product design input. The human factors research could have made a valuable contribution to the design if it had been done at the correct phase of the project. Additionally, the human factors "analyst" was unable to read engineering drawings and communicate with design engineers. It became evident that the company should have checked the work experience and educational background of the person and not relied on this person's title.

4.2.3 Specialists versus generalists

Once you have analysed your project and decided what category of human factors professional you need, you must decide if you need a specialist or generalist.

Human factors is a very broad discipline. Some professionals are specialists in one or more fields of human factors. These specializations include human response to noise, anthropometry, human cognitive abilities, or group dynamics. Other human factors professionals are generalists. We consider ourselves generalists. This book preparation required the skills of a generalist. A generalist understands research techniques, has a general understanding of the capabilities and limits of humans, and knows where to seek specific data on humans. Human factors generalists rely on human factors specialists to advance specific areas of human factors knowledge. We would not, for instance, attempt to design a target tracking system in a supersonic aircraft without consulting specialists in control dynamics and feedback systems.

When selecting a human factors professional, be aware of the difference between a specialist and a generalist. If the system requires consideration of the human throughout the entire design process, then select a generalist. Select the human factors professional who has had total systems planning and development experience, particularly with the same type of system under development. This person will be able to work with the design staff to bring timely and cost-effective answers to human-machine problems.

If the system must address unique human interface problems, it may be necessary to add a specialist to the human factors team. Designing a space suit, for instance, will require a human factors anthropologist who specializes in measuring and fitting the size, range of motion, and strength of the astronaut population. Other human factors specialists are research oriented. For example, robots remotely controlled from Earth will explore Mars. The transmission of a radio signal between Earth and Mars can take 15 min, making feedback and control very difficult. Human factors specialists will test and evaluate human performance to design the best possible workstations for controlling these robots.

4.3 Locating Human Factors Professionals

We know of companies that have spent several years looking for full-time human factors professionals. At present the human factors profession contains a limited number of people. In the future your search to locate human factors professionals will become easier because the industry is growing rapidly and colleges and universities are graduating more human factors people.

The Human Factors Society and other professional organizations are sources of human factors professionals. These are listed in Chapter 12. Most have placement information and can supply you with the names of individuals and companies providing human factors services.

Colleges and universities are excellent sources for human factors help. The placement office can supply you with people looking for employment in human factors. These may be students graduating with degrees in human factors or they may be people updating their skills in human factors. By contacting either the engineering or psychology departments an employer may locate human factors help. A professor may be available as a consultant or they may have students available to help with human factors projects. Human factors professors can also give you information on former students and people they have worked with.

If you have contracts with the government, contract administrators often have lists of human factors professionals or companies. A company we were working with obtained a very large contract with the Army. Our company did not have enough human factors personnel to complete the contract on schedule, so the Army recommended a human factors consulting company. This approach to finding human factors help can apply to subcontracts within industry as well.

4.4   Hiring a Human Factors Professional

You need to carefully select the human factors professionals that will contribute the most toward improving your design. To do this there are evaluation techniques for finding a competent professional. Evaluate the candidate through references from previous clients and employers. Ask questions.

Find out what type of work the person did for the client. Did the human factors professional act as a generalist responsible for all or nearly all interfaces between the hardware and operators and maintainers? Or instead, did the person concentrate on one or two specific human factors problems?

Find out if the human factors professional acted independently and met the client's contractual needs. Did the human factors professional supervise others and prepare progress and budget reports? Depending on the answers, you must budget time for supervising the human factors professional.

Finally, ask about the quality of the work produced for the client. Did the human factors professional meet the deadlines within schedule and budget? Were the products useful and incorporated into the system design or were they merely filed? How well did the person work with others responsible for the product design? A valuable human factors professional is willing to share information and teach the client.

Before interviewing a human factors professional, quickly review design standards and criteria for your product that relates to human factors. This may include standards from professional societies such as the Society of Automotive Engineers, or military human factors design standards. Jot down the standards and have them in front of you when you interview the person. Human factors professionals should have a working knowledge of human factors requirements and feel that they are capable of helping you meet them. If your product has no design standards, human factors professionals may locate standards that can be used for a design goal. This may give you an edge on the competition. A furniture manufacturer might advertise that their chairs are "ergonomically" designed to fit comfortably 95 percent of the adult population.

The human factors professional that communicates well can add strength to a proposal. Evaluate oral communication and presentation in your interview. Ask to see something the human factors professional has written, both formal and informal. Often others edit a report and it doesn't reflect the communication skills of the person you are interviewing for the job.

If you want the human factors professional to work with designers, ask to see samples of drawings, sketches, models, or other graphics. A busy designer often does not have time to interpret volumes of text. Communications will be more efficient if it is in a form directly usable by the designer.

If you are hiring a human factors person for a subcontract, ask for a proposal. This can be a one-page proposal or a long formal proposal depending on the project.

Make certain that the human factors professionals truly understand your problems and are realistic about meeting your needs. Do not accept statements like "no problem," "we'll take care of it," or "your problem is so complex only an expert can understand the solution." You understand your product and your needs better than anyone else. If they cannot communicate to you, how will they be able to communicate with your design staff? Ask for a proposal with deliverables or results that will meet your needs on time. The results should be measurable.

Check to see if the human factors person is professionally competent. The human factors field is growing in two ways. Research is expanding what we know about humans and technology is changing the ways humans interface with their environment. The human factors professional must keep current with both research and technology advances. Professional status is kept current through contacts with colleagues, colleges and universities, workshops, and field experiences. A human factors professional who works with other companies may have valuable insights on how other companies solve problems.

Many employers attempting to locate and hire human factors professionals use directories developed by professional associations, commercial publishing companies, and government agencies. We have found from personal experience that most publishers of these directories develop them either by buying mailing lists or using their own member rosters to send out questionnaires. These questionnaires ask about degrees, areas of interest, and sometimes experience. They often require the professional to pay a listing fee. Be very careful in using these lists to determine if an individual is a qualified human factors analyst or human factors engineer. Even professional associations cannot validate their membership and there is no professional licensing for the field of human factors. These directories can locate individuals and company names for you, but you must validate the degrees and work experience yourself. Suggested guidelines for hiring a human factors professional appear in Fig. 4.2.

Determine which human factors candidates to interview
1. Analyze your project to determine if you need an applied human factors professional or if you need a research human factors professional.
2. Determine if you need a human factors specialist or a generalist.
3. Select the applicants to interview based on their educational background and work history that meet your requirements for items 1 and 2 above.

Before the interviews
1. Using the applicant's resume, contact previous clients and ask:
   (a) Did they meet the client's requirements?
   (b) Did they meet the deadlines and stay within the budget?
   (c) Were the applicant's suggestions included in the final design (if hiring an applied human factors professional)?
2. Review human factors standards for your project.

During the interviews
1. Ask the applicant for suggestions (specific examples of what they will do for your project).
2. Evaluate their presentation skills. Can they present their ideas clearly to your design team?
3. Ask to see a written report prepared by the human factors professional.
4. If the professional will work directly with engineers or designers, ask to see examples of their drawings, sketches, models or other graphics.
5. Ask for a written proposal with deliverables and deadlines.

Post interviews
1. Review proposal—look for specific answers and clear solutions to your human factors design problems. The entire design team must be able to understand and carry out the solutions presented by the candidate.
2. Select candidate and agree on a contract, proposal, or job description.

Figure 4.2   A checksheet for hiring a human factors professional.

5

How Do I Manage
a Human Factors Program?

5.1 Checklist for a Human Factors Project

This chapter will help you do a better job managing and working
with a human factors professional. Human factors is an important
consideration in the design and development of systems. The human is a major component of a design and is as important as any
mechanical component. If the human component fails, the system is
usually doomed to failure. As systems become more complex, human factors is increasingly important in the design. Now, even the
home has sophisticated equipment that is inoperable unless human
factors is incorporated into the design process. Therefore, if you are
involved in or responsible for a system design, it is important for
you to understand how to manage a human factors program.

Figure 5.1 is a checklist to help you manage your human factors
effort. The following paragraphs explain the checklist items.

1. Project definition

Defining the job is the most important aspect of managing the human factors professional.

a. Goals

First, have a clear definition of the project goals. You are in
the best position to understand your requirements. If you have
a government contract, human factors requirements will be de-

MANAGEMENT CHECKLIST FOR A HUMAN FACTORS PROJECT

1. Project Definition
   ___a. Goals.
      ___do they solve the problem
      ___do the goals cover applicable human factors design standards
      ___do the goals address potential product liability problems
   ___b. Techniques
      ___define problem in greater detail
      ___research and analyse possible solutions
      ___test and evaluate problem solutions
      ___implement solutions
      ___validate solutions
   ___c. Schedule Activities
   ___d. Products
      ___final and more detailed program plan
      ___progress reports (verbal, written, and formal presentation)
      ___tests plans and test results
      ___final product (reports, drawings, mockup, computer program, etc.)
      ___follow up evaluation report
   ___e. Compensation
      ___amount
      ___schedule of payment
      ___contingencies of payment
   ___f. Miscellaneous Expenses
      ___travel
      ___test subjects
      ___reference material
      ___computer facilities
      ___copying
      ___mockup and drawing materials
      ___communications and mailing expenses
   ___g. Working Arrangement
      ___independent contractor or employee of your company
      ___work location
      ___communications (letter, telephone, fax, telecommunications, etc.)
   ___h. Document

2. Support The Job
   ___a. Introductions
      ___your management
      ___your customer
      ___personnel that will work directly with the human factors
      professional.
   ___b. Provide Facilities
   ___c. Meet Professionals Needs
      ___review reports and ask for feedback
      ___arrange meetings if necessary (with customer, with management, etc.)
      ___insure the all departments are cooperating with the human factors
      professional (this includes engineering, contracts administration and
      accounting, travel, library, and reproduction departments)
      ___route reports, memos, letters concerning the project to human factors
      professional

3. Manage The Job
   ___a. Arrange Frequent Meetings
   ___b. Request and Review Products
      ___final program plan
      ___progress reports
      ___final report
      ___follow up report

4. After The Job
   ___a. Reevaluate
      ___Was the product useful
      ___Did the final report receive a positive response
      ___Did the customer respond positively

Figure 5.1 Checklist for managing a human factors project.

fined in the contract. If you have a commercial contract, you have an idea of the industry and government codes relating to human factors that you must meet. Consider product liability. Many customers are realizing that product designers and manufacturers can be held accountable for accidents and illnesses resulting from the use of their products. Many problems can be avoided by setting goals to reduce human error and human exposure to potential hazards.

The human factors professional can help with setting program goals. A telecommunications company hired us to help their human factors effort on an Air Force contract. Their problem was that the contract called for using government human factors standards that did not apply to telecommunications workstations. We were aware of newly developed government standards for workstations. These standards were so new that they were not required in the contract. We were able to help the company set their human factors goals by including the new standards. The goals met the contract requirements and satisfied the customer.

b. Techniques

Once you and the human factors professional have agreed on your goals, have the professional outline how they intend to meet the goals. The checklist shows five human factors program stages. There should be a technique or method for each stage. You should understand the process that the human factors professional intends to use. By understanding the human factors professional's approach you can evaluate the approach and change it if necessary. Sometimes human factors professionals are so involved in perfecting analytical techniques that they forget your limits of time and money. For example, doing extensive human performance tests to evaluate different design options would not make sense if the engineering department had already released drawings to fabricate the product.

c. Schedule activities

Schedule human factors activities. Many times the human factors professional will be unfamiliar with the process of design and manufacturing. It is important to the program to provide timely human factors inputs.

d. Products

Define the human factors effort carefully. The checklist has typical products and deliverables. The following is a brief description of these products and what they should include.

Program plan: First, you may want the human factors professional to study your problem further and give you a more detailed plan to resolve the problem. You may be able to implement the plan and resolve the problem without further help from a human factors professional.

Progress reports: You will want formal progress reports from the human factors professional. You need them to include in your own reports to management and to your customer. The report should describe work the human factors professional has done, problems incurred, and planned activities for the next reporting period. Progress reports should allow you to compare what was completed with what was promised.

Test plans and test results: Much of human factors work is accomplished through testing and you may want the human factors professional to give you test plans and test results. There are generally two types of tests: development tests and validation tests. Development tests are tests conducted during product design to find information about humans. For example, you may need to know where to put a handle so all operators can reach it. Human factors validation tests are usually conducted at the end of a design program to "prove" people can operate and maintain the product. A validation test plan is important because the test must cover all potential problem areas. A validation test report is valuable because it will tell you how successful the human factors effort was and what problems still exist.

Final product: Select the final output of the human factors program. The final output may be a set of recommended operator workstation drawings, a mockup or model of the station, or a report outlining problems in a design and recommendations for eliminating them. Be certain the final output is responsive to the program goals and is useful to the designers.

Follow-up evaluation: It is a good idea to request a follow up study and report. The human factors recommendations may take time to carry out and the results will not be apparent. You may want the human factors professional to come back at a later time to study your product and give you an evaluation.

e. Compensation

Decide the compensation for the effort. The amount you pay for the human factors specialist depends on many items including your budget. Consider the consequences of not having a human factors effort. They may include loss in sales, product redesign, or product liability costs. Compare the cost of a human factors professional with the design or research specialists' salaries in your company. Consider the differences between what an independent contractor costs your company and the cost of a full-time employee (medical benefits, retirement plan, etc.). Decide the payment schedule. Some companies will pay simply on the hours worked. You may have better control if you pay based on the product delivered.

f. Miscellaneous expenses

List miscellaneous expenses and decide who will pay them. These items can make an impact on your project budget. These additional cost items may be considered overhead by a human factors company or a human factors professional may expect your company to provide them. Each item on the list should be discussed and agreed on before beginning the project.

g. Working arrangement

Consider the working arrangement in the job definition. Decide if the human factors professional should work as an independent contractor or instead, should work directly for your company. Both arrangements have advantages. If your organization foresees a sustaining need for a human factors specialist, then maybe you should encourage direct employment. If this is not possible, an individual in your company could work with an outside expert until they are "up to speed" to take over project responsibility. If you see gaps in your need for human factors expertise it may be better to hire a professional outside your company. When the project ends the human factors professional can go on to other jobs. You would not be responsible to find work for an unnecessary specialist on your staff.

Another important aspect to consider is where the human factors professional will work. We have found a design and development project that involves a large staff such as engineers, draftspersons, and machinists requires some human factors on-site exposure. On-site presence will help engineers integrate human factors data into the design. An engineer designing a product which interfaces with a human will need information on the size, strength, and speed of humans. For instance, a tall engineer may estimate that all adults in a nuclear power plant can reach an emergency switch 6-½ ft from the ground. Nearly 25 percent of the population will not be able to reach the switch. If the engineer sees the human factors professional on-site, they will realize it is simple to ask for the proper data.

Other human factors tasks such as research and analysis are better conducted uninterrupted at the professional's facilities. If the human factors professional works at a remote site, plan the best means to communicate. This includes communications for regular information exchange and communications for resolving unexpected problems.

h. Document

Document and sign the agreement, or statement of work, and have the human factors professional sign it. The agreement will make everyone's job easier. Figure 5.2 is an example of a statement of work.

2. Support the job

a. Introductions

One of the more important things you can do to make sure the human factors effort is a success is to help the professional establish a communications network. Introduce the human factors professional to all people they will work with. Managers sometimes do not understand the human factors professional's role. One company asked us to design their product: a cab for coal mining equipment. It was important for the cab to fit the operator but we are not trained to design the suspension sys-

STATEMENT OF WORK
PURCHASE ORDER NUMBER _____

MANAGER _____A. Manager_____
CONTRACTOR ___Tillman Ergonomics Company, Inc.___
TOTAL COST _____$200,000_____

WORK TO BE DONE:

Analyze mission functions and recommend procedures and console
arrangement to reduce crew size from 3 to 2. All recommendations will
meet applicable human factors standards.

DELIVERABLES:
ITEM   DUE DATE   COST
                                                     (If Applicable)

1. Report defining duty assignments   3/92
for 2 person crew.
2. Drawings showing recommended   6/92
control console with control and
display arrangement.
3. Report detailing: 12/92
        Operating procedures
        Task times
        Control/Display/Software
        specifications

SIGNATURES:

        CONTRACTOR _____ DATE _____
        MANAGER   _____ DATE _____
        APPROVAL   _____ DATE _____

Figure 5.2   Example statement of work for hiring human factors professionals.

tem and hydraulic controls. We explained to the manager we
would work with the engineers and provide information including the cab envelope dimensions. To expect us to design
the entire cab would waste both our talents and the talents of
the engineering staff. Engineers are specialists in mechanics,
hydraulics, and electronics. The human factors professional

who works with engineers should be thought of as a resource specialist for the human component in the system.

b. Provide facilities

Provide the human factors professional with working facilities required by your agreement. You don't want valuable time wasted while they look for a desk and chair.

c. Meet professional's needs

Review all the human factors reports and studies to keep an understanding of the progress and problems that are developing. Ask the human factors professional for feedback on problems and ways you can help. Keep the human factors professional informed of management decisions that concern them. Convince all departments to cooperate with the human factors professional. Help the human factors professional deal with "in-house" procedures such as meeting locations, times, and report and memo publications.

3. Manage the job

a. Arrange frequent meetings

Managing the human factors professional should be budgeted into your time. You will need frequent informal meetings. The amount of time required to manage them depends on the project and the professional's competence. We estimate 5 percent of the human factors time budget to manage the human factors professionals. You may want to budget considerably more time.

b. Request and review products

Understand what the human factors professional is doing. Make sure you are getting what you agreed to in the contract. Unfortunately we have seen human factors professionals live for years off promises. They convince their employer that what they are doing is too complex to explain, but the results will be wonderful. You must require the human factors professional to put something that will improve the design on your desk when the project ends. It would even be better if you obtained something every month or every week. This will prevent any unpleasant surprises at the end of the contract.

The items under 3b on the checklist are products you should expect to receive from the professional. The most important feature of these reports is that they contain information to improve your product. Make sure each report does this.

4. After the job

a. Reevaluate

Were all contract requirements met on time? What went right and what did not? Provide funds in the budget for the human factors professional to contribute to the assessment. Your notes may help you decide how you are going to handle your next human factors project. If you do not have a permanent human factors staff maybe you will think it is worthwhile to establish one.

Another helpful experience is to review the job sometime after its completion. Find out if the recommendations were carried out and if they were successful in achieving the goals. This review will give both you and the human factors analyst valuable information. There is often a long process between an initial idea or concept and the final implementation.

5.2   Ways to Obtain Additional Human Factors Resources

When managing human factors professionals you should ask them to explain exactly what they are doing and the rationale and theory behind it. Human factors professionals should be happy to share what they know and help you to learn more about the profession. However, you may feel more comfortable to have your own resources and references handy when managing the job.

Figure 5.3 lists these additional resources according to the types of information you may need. Specific names and addresses for these resources are in Part III of this book.

INFORMATION REQUIRED	POSSIBLE INFORMATION SOURCES
General Human Factors Design Guidelines	Human factors handbooks and design guides published by commercial publishers Company design guides and handbooks U.S. Government publications (Department of Defense, National Aeronautics and Space Administration, Department of Labor)
Human Factors Design Standards For Specific Systems	U.S. Government publications (Department of Defense, National Aeronautics and Space Administration, Department of Labor) Industrial standards
Specific Human Factors Data (human performance data, environmental effects on humans, strength and size of a people in a specific occupation, etc.)	Research periodicals published by human factors professional societies Human factors professionals Other professionals (physicians, physiologists, psychologists, anthropologists) Colleges and universities Research reports and papers (accessible through libraries or specialized report distribution centers)
Names Of Human Factors Professionals	Human factors professional societies Telephone book (under human factors or safety) Colleges and universities
Information On Human Factors Analytical Techniques and Tools	Human factors textbooks (college bookstores) U.S. Government (Department of Defense) Human factors professionals
Information On Human Factors Training Courses	Colleges and universities Human factors professional societies

Figure 5.3   Human factors information sources.

Specific Answers to Help Improve Product Designs

6

How Do I Use Human Factors in the Design Process?

6.1 The Design Process

Once you decide to apply human factors to a design you will need analysis techniques to apply it to the design process. We will review human factors techniques and methods to analyze designs. Our purpose here is to explain briefly how to conduct the analysis. These techniques provide design recommendations for use in the design of either products or entire systems. Part III lists references for the techniques.

Designs progress through stages and human factors techniques and methods should be included in all stages. Figure 6.1 lists design process stages and human factors analyses for each stage. To illustrate how to conduct human factors analysis, we will follow a design through its development. We use the following statement of work as an example.

Statement of work

Design space vehicle crew accommodations for a 1-year mission to Mars and back. The design must support the crew members both physically and mentally. The support includes nutritional facilities, workstations, exercise facilities, rest areas, and minimum living volume.

6.2 Systems Analysis in the Preliminary Design Stage

Human factors in the design process begins with systems analysis techniques. The human factors goal in systems analysis is to iden-

DESIGN STAGE	HUMAN FACTORS ANALYSIS
Preliminary Design	Systems Analysis Identify System Requirements and Constraints Define Functions Allocate Functions
Detail Design and Development	Task Analysis
Final Test and Evaluation	Final Test Prepare Test Plans Monitor Testing Analyze Test Results

Figure 6.1   Human factors analysis at each design stage.

tify the people in the system: Who is going to operate and maintain the system, what will their duties be, what environment will they work in, and what demands and stresses will they work under? Systems analysis has three steps:

1. Identify the system requirements and constraints related to the human.
2. Define the essential system functions, the function sequence, and the environment (cold, hot, dark, etc.).
3. Select the people, hardware, and software to execute the functions.

The first step in systems analysis is to identify the requirements and constraints related to the human in the system. In a government contract these human factors requirements and constraints are easy to find. Sometimes they are under a separate human factors heading in the contract. A commercial product, however may have less definitive human factors requirements and constraints. The project team may have to make its own requirements and constraints based on what is known about the customers. Industry or government standards may dictate other requirements and constraints.

In our Mars mission example (Section 6.1) the contract may have the following requirements and constraints for the spaceship crew:

1. The spaceship shall accommodate 4 crewmembers for 12 months to Mars and back.

2. Each crewmember shall have a private sleeping quarter.

3. The galley shall accommodate all crewmembers simultaneously.

4. The crew shall be able to prepare for activities outside the spaceship (put on space suit, etc.) in 1 hour.

The designer must create a design that meets all the requirements and constraints. You can help the designer meet the human factors requirements and constraints with human factors design standards and guidelines. Chapters 7 and 8 will discuss standards and guidelines in more detail. Research these standards and guidelines to find information related to the human factors requirements and constraints.

NASA has human factors standards for the habitat volume necessary to support crews for extended periods. This information will help the designer meet the first Mars mission requirement listed above. Human factors information also exists to help the designers create crew quarters for requirement number 2. You should research and consolidate all information about humans to help the designer meet the human factors requirements and constraints. Make the human factors information easy for a designer to apply. Use graphic and tabular formats whenever you can. This information package is called a design data book. Deliver the design data book to the appropriate designer.

The second analysis step is to define the system functions. An excellent way to define system functions is to start with the final system goal and work backwards through the steps. Government contracts usually specify system functions. On other programs you may have to research similar systems to determine system functions. NASA would specify the system functions in our Mars mission example.

One simple technique for describing the system functions is the functional flow block diagram. Figure 6.2 shows a functional flow block diagram for our Mars mission. Each block is in the proper se-

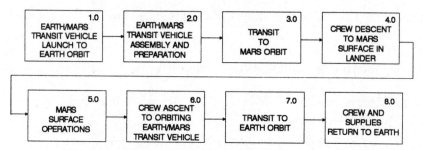

Figure 6.2   Example of a functional flow-block diagram for a Mars mission.

quence and shows a short system function description. The function description tells what is done, not how to do it.

Another way to show system functions is the decision/action diagram. A decision/action diagram interprets all functions as actions and yes-no decisions. Figure 6.3 is a decision/action diagram. The

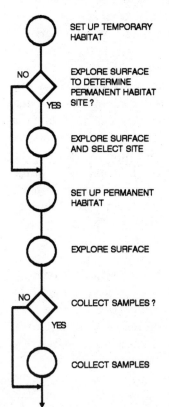

SET UP TEMPORARY HABITAT

NO   EXPLORE SURFACE TO DETERMINE PERMANENT HABITAT YES   SITE?

EXPLORE SURFACE AND SELECT SITE

SET UP PERMANENT HABITAT

EXPLORE SURFACE

NO   COLLECT SAMPLES?

YES

COLLECT SAMPLES

Figure 6.3   Example of a decision/action diagram for a Mars mission functional flow block for 5.0 Mars surface operations.

decision/action diagram is a graphic way to show detailed function sequence. You can combine it with the functional flow block diagram technique, by using the decision/action diagram to describe the details of each functional block.

The third and final step in systems analysis is to select the people, the hardware, and the software to do the system function. The term for this is functional allocation. Breaking functions into steps helps to determine how to allocate the functions. Human factors play an important role in functional allocation. Human factors standards and guidelines have information on what humans can do and what jobs are best done by software or hardware systems. Other things to consider in functional allocation include cost, technical feasibility, and reliability.

In summary, systems analysis produces the following results:

1. *Human factors design data book:* A report containing human factors design requirements, design standards, and guidelines for meeting these requirements.

2. *Function description:* The sequence of functions the system will do.

3. *Personnel requirements:* Information about people required to operate and maintain the system; a preliminary description of operator and maintainer duties.

These products form the next human factors step in the design process: task analysis.

6.3 Task Analysis in Detail Design and Development Stages

Human factors contributes extensively with task analysis methods during the detail design stage. The human factors task analysis goals are to provide detailed human factors information to designers and to determine system operating procedures. Task analysis information helps you determine designs, training, and operating instructions to help the user do the task safely and efficiently. Task analysis helps define the personnel qualifications to operate and maintain the system. Task analysis also helps you determine how to test the system to be certain people can successfully operate and maintain it.

A task analysis is a detailed description of what humans in the system are doing. The task analysis defines the following:

1. Equipment used to do each task

2. Stimulus initiating the task

3. Human response

4. Task feedback

5. Task performance criteria

6. The task environment

There are several techniques for you to use to do task analysis. We will discuss three briefly here: operational sequence diagrams, time lines, time and motion analysis.

1. *Operational sequence diagram:*   The operational sequence diagram (OSD) technique shows the coordination between people and machines. Figure 6.4 is an OSD worksheet. It shows coordinated op-

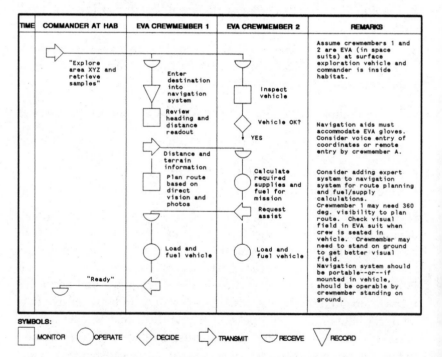

Figure 6.4   Example of an operational sequence diagram which shows analysis of tasks to prepare a Mars surface exploration vehicle.

erations among three crewmembers on the Mars surface. Two crewmembers are outside the living habitat preparing a land exploration vehicle. The third crewmember is inside the habitat. We divided the OSD into three columns: one column for each person in the system. The six symbols in Fig. 6.4 define the people and equipment operations. Time flows from OSD top to the bottom. The symbols show who is doing what. Because the OSD is a graphic presentation it helps you identify potential problems. For example, the OSD may show one operator receiving more messages simultaneously than they are able to respond to. Or the OSD may show all the symbols in one column suggesting operators may be unnecessary and should be eliminated. The last OSD column is for observations made during the task analysis process. The analyst can make notes about potential problem areas, design ideas to improve the task, or areas to research.

2. *Time lines:* Time lines are useful for analyzing time-critical tasks: handling emergency telephone calls, duty assignment during a submarine maneuver, getting a car out of the pit and back into the race. Figure 6.5 shows a portion of a time line. This time line is

| TASK CONSTRUCT HABITAT ON MARS | PERSONNEL SYSTEM | 3 EVA CREWMEMBERS (A, B,C) GEODESIC DOME STRUCTURE | NAME TILLMAN DATE REV | PAGE 1 OF 3 |

Figure 6.5   Example of a time-line worksheet.

for three crew members (A, B, and C) constructing a habitat on the surface of Mars. The top of the sheet shows time intervals. The left column shows the tasks that should be done. The bars to the right of the tasks show how long each task will take and who will be doing the task. If you need to complete several tasks in a limited period, the time line helps distribute tasks so all personnel are busy. There are several computer programs to help with time line analysis. The programs select the task assignments to minimize the time.

3. *Time and motion analysis:* Time and motion analysis is perhaps the oldest task analysis method. F. Gilbreth developed the technique in 1911. Time and motion analysis looks at the task's microelements such as hands, fingers, and eye movements. To use time and motion analysis you assign each microelement a certain time. You can use the analysis to minimize repetitive task times. This method can help design a computer board assembly workstation. Time and motion analysis can show the best place to locate microchips, a microscope, or cleaning brush to reduce assemblers search and reach times.

Noncritical tasks are tasks that cause no potential harm to the user, equipment, or system. You analyze noncritical tasks by estimating times and potential problems. Tasks critical to successful systems operations should be analysed by task simulation. You may need to use mockups or simulators with test subjects to get information about the task. Chapter 7 discusses mockups and simulators in more detail.

Task analyses produce the following results:

1. *Design inputs:* Tasks analysis shows ways for improving the design so people can be safer and more productive. For example, a task analysis may show that an operator must read a dial under both bright and dim lighting. Human factors references give standards for number size, dial lighting, and dial placement. After you determine the design inputs you should transmit your findings directly to the engineer working on the project.

2. *Duty assignment:* Task analysis results will help you assign tasks to personnel and make the most efficient use of their time.

3. *Procedure development:* Analyzing tasks will help determine

the most efficient task sequence. Task sequence should be placed in instruction and operating manuals.

4. *Training requirements:*   Analyzing tasks help determine duties and skills required by each person in a system. This information can be built into personnel selection criteria and into training programs.

5. *Information for test plans:*   Task analysis and functional analysis help identify critical tasks and tasks humans may fail to do. You should test these tasks in the final product design.

6.4   Human Performance Evaluation in Final Test and Evaluation

Human factors tests the final design to verify if humans are able to successfully operate and maintain the finished product. A validation test failure may suggest a need for an improved design, personnel selection criteria, training, or system operating and maintenance procedures. Human factors participates in the final product test two ways: test plan preparation and test-results interpretation.

1. *Test plan preparation:*   Human factors contributes to test plan preparation in two ways. First, human factors identifies what to test. You can identify tasks to test using systems and task analysis methods described earlier. Your test plan should concentrate on the critical tasks you found. Failure of critical tasks could harm the user, equipment, or system. You must also test difficult tasks. In our Mars mission example (Section 6.1), we may need to confirm that crewmembers can hear a fire alarm warning while they are close to noisy machinery.

   Second, human factors contributes to test plan preparation by specifying the test subjects, what the test subjects will do and how to measure their performance. You can write test procedures by using your task analysis findings described earlier. For example, we worked on military systems requiring crewmembers to accomplish tasks (such as preparing a weapon) in a specified time. We analyzed the tasks and experimented with subjects and mockups. We established the most efficient procedure for doing the task and incorporated the procedure in the final test plan. We found limited space caused the largest crewmember to take longer. Because large crewmembers took the longest time, we tested the worst possible conditions, we specified that the test subjects be 95th percentile in stature.

2. *Test results interpretation:*   Human factors monitors final testing and helps interpret the results. Human error can cause test failure. By knowing how the human should perform you can identify errors. Analysts can help redesign the system to reduce human error and successfully complete a retest. The human factors analyst can determine other valuable information by carefully observing test subjects. We have observed a test crew successfully loading a weapon in a timed validation test. The crew was not following the procedures in the technical manual. Instead, they had developed a faster procedure. We revised the technical manuals with the improved procedure and met our performance requirements.

   In summary, human factors should do the following during the final product testing stage: (1) Identify the critical and difficult human tasks, (2) specify the task procedures, the test subjects, and the method for measuring human performance, and (3) monitor the test and interpret the results.

6.5 Summary

Many techniques exist to meet design requirements. Three basic human factors steps in the design process are:

1. *Functional analysis and allocation:*   Human factors must analyze and allocate system functions to determine the duties and demands on people in the system. If people in the system are overburdened they will make mistakes, become fatigued, and cause system failure. If people are underburdened they will become inefficient, bored, and distracted.

2. *Task analysis:*   Analyze tasks after you have allocated functions to humans. The environment should be safe and conducive to work, hardware should be simple to operate, procedures should be understandable and easy to follow, and training and personnel selection should prepare the worker to do the job correctly. Task analysis examines the demands of each task and helps define ways to better support the human.

3. *Final testing:* Final tests validate that humans can do tasks safely and efficiently. Human factors participates in final testing to make sure the tests realistically represent the demands on human performance. Human factors also helps interpret results and recommend ways to improve human performance.

We have briefly explained specific human factors techniques. You can find more detail on these techniques and others in the references in Part III. The next chapter gives information on tools used during human factors participation in the design process.

7

How Do I Use
Human Factors Tools?

7.1 Human Factors Tools Are Available

Understanding human factors tools and how to use them will help
you to better resolve human factors problems. You may want to apply these human factors tools directly to your design, or, hire a human factors professional specializing in working with these tools.
This chapter describes tools available to the human factors professional or anyone wishing to do human factors work.

There are three types of human factors tools:

1. Human size and performance design guidelines

2. Measurement tools—tools to measure human performance and
working environments

3. Analytical tools

The following material describes each tool and how to use them.
A figure summarizes information at the beginning of the following
sections. If you are trying to resolve a particular human factors
problem, look in the "tool use" column of the figure. The left column
describes and names the tool. After determining which tool you
need for your project, you can locate details on each tool in the references listed in Part III.

7.2 Design Guidelines

During hardware or software design, human factors professionals
consistently use design guidelines and information on human size

and performance capabilities. The information in design guidelines is based on scientific research. These design guidelines are easy to read and apply. Often the information is in a tabular or graph form. Figure 7.1 summarizes design guideline tools.

TOOL	DESCRIPTION	USE
Function Allocation Guides	Lists comparing human capabilities with machine capabilities	Selection of number and type of people to operate a system Allocation of jobs to humans or machines
Workplace Arrangement Guides	Physical dimensions of the best areas to locate controls and displays in a workstation	Workstation preliminary design and layout
Control and Display Selection Guides	Guidelines for selection of controls and displays based on the functions that they must perform	Selection of controls and displays for operator workstations
Control and Display Design Guides	Dimensions and operating force limits for controls and displays (dimensions and force limits are usually defined as maximum, minimum, and preferred)	Control and display detail design
Software Design Guides	Lists of rules detailing how software should interact with the human	Design of computer software interface with the operator
Human Size and Capability Data	Tables and charts showing: • Physical Dimensions • Movement Range • Human Cognitive Skills • Range of Human Perception	Hardware design to fit human size and abilities
Environmental Design Guides	Tables showing the range of environmental conditions for human comfort and productivity Environmental parameters include: • Light • Temperature • Humidity • Ventilation • Noise • Vibration	Workstation location and design of environmental control systems such as heating, lighting, cooling, insulation, and ventilation

Figure 7.1   Human factors design guideline tools.

7.2.1 Functional allocation guides

With the sophistication of automation and robotics, many tasks can be allocated to machines instead of humans. Careful task allocation is particularly important in dangerous environments, such as combat situations or a space environment. The human factors profession participates in deciding which functions should be done by machines and which by humans. There are lists comparing human and machine capabilities to help designers and human factors professionals make these decisions. One of the earliest and most famous lists is the Fitts list. Figure 7.2 is a copy of Dr. Paul Fitt's list de-

HUMANS EXCEL IN	MACHINES EXCEL IN
Detection of certain forms of very low energy levels	Monitoring (both humans and machines)
Sensitivity to an extremely wide variety of stimuli	Performing routine, repetitive, and very precise operations
Perceiving patterns and making gereralizations about them	Responding very quickly to control signals
Ability to store information for long periods and recalling relevant facts at appropriate moments	Storing and recalling large amounts of information in short time periods
Ability to exercise judgement where events cannot be completely defined	Performing complex and rapid computations with high accuracy
Improvising and adopting flexible procedures	Sensitivity to stimuli beyond the range of human sensitivity (infrared, radio waves, etc.)
Ability to react to unexpected, low probability events	Doing many different things at one time
Applying originality to solving problems (i.e., alternate solutions)	Exerting large forces smoothly and precisely
Ability to profit from experience and alter course of action	Insensitivity to extraneous factors
Ability to perform fine manipulation, expecially when misalignment is not expected	Ability to repeat operations very rapidly, continously and precisely the same way over a long period
Ability to continue to perform when overloaded	Operating in environments hostile or intolerable to humans
Ability to reason inductively	Deductive processes

Figure 7.2 Functional allocation guide comparing human and machine capabilities. (*Adapted from DOD-HDBK-763, "Human Engineering Procedures Guide," Department of Defense, February 1987.*)

veloped for the U.S. Air Force in the 1950s. There are several updates, and the list is still used.

7.2.2 Workspace arrangement guides

The system design requires a preliminary workstation layout. Hand controls, foot controls, and visual displays should be placed where they can be reached and seen. There are charts to help locate controls and displays in both standing and seated positions. They provide the boundaries for locating controls and displays. There are even charts to help locate controls and displays in microgravity conditions of space (where people and objects float). Figure 7.3 is an example showing design guidelines for hand control location (on Earth).

Control location guides are based on peoples' reach capabilities. The guides consider the full population size range. The outer boundaries are within a small person's reach limits. The inner boundaries do not restrict or impair a large person's body movement.

When using the control and display location guides it is necessary to have a reference point. Two of the most common reference points are the eyes and the seat reference point. Eyes are an important reference point because operators must be positioned so that they can see a display. The seat reference point (SRP) is used for seated workstations. The SRP is at the intersection of the seat pan cushion and the seat back cushion (depressed with body weight). Charts show the SRP relationship to the eye and the control reach envelope.

7.2.3 Control and display selection guides

As the design process continues, the designer selects the best controls and displays for the operator. Depending on the task and work environments, certain controls and displays are better than others. For example, an indicator light in a control panel is a good way to warn an operator who must remain at a single workstation (such as an automobile driver). A panel light will not work for a chemical plant operator who may be at many different workstations. Here, an audio signal may be the best warning display. Human factors

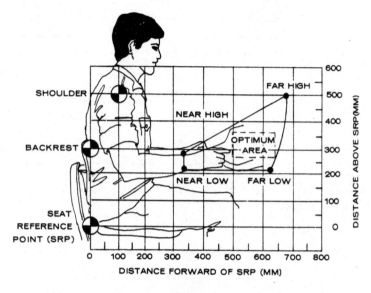

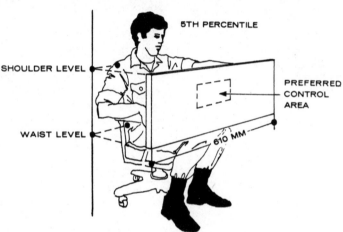

Figure 7.3 Workspace arrangement guide showing optimum manual control locations for seated operations. (*Reproduced from MIL-HDBK-759A, "Human Engineering Guide for Army Material," Department of Defense, 1981.*)

guides can help make these selections. Figure 7.4 shows an example of a control selection guide.

7.2.4 Control and display design guides

Human factors has detailed control and display design guidelines. Figure 7.5 shows an example of a human factors control design guideline.

Advantages	Disadvantages
a. Joystick	
Can be used comfortably with minimum arm fatigue	Slower than a light pen for simple input
Does not cover parts of screen in use	Must be attached, but not to the display
Expansion or contraction of cursor movement is possible	Unless there is a large joystick, an inadequate control/display ratio will result for positional control
	The displacement of the stick controls both the direction and the speed of cursor movement
	Difficult to use for free-hand graphic input
	Not good for option selection
b. Four arrow cursor control	
Allows accurate positioning of the cursor	Should not be used for free-hand graphics
May provide positive transfer and advantages associated with touch typing	
Allows for nondestructive movement of the cursor	
Requires little or no training	
c. Light pen	
Fast for simple input	May not feel natural to user, like a real pen or pencil
Good for tracking moving objects	May lack precision because of the aperture, distance from the CRT screen surface, and parallax
Minimal perceptual motor skill needed	
Good for gross drawing	Contact with the computer may be lost unintentioanlly
Efficient for successful multiple selection	Frequently required simultaneous button depression may cause slippage and inaccuracy
User does not have to scan to find a cursor somewhere on the screen	Must be attached to terminal, which may be inconvenient
May be adaptable to bar coding	Glare problem if pen tilted to reduce arm fatigue
	Fatiguing if pen is held perpendicular to work surface
	If pointed to dark area, may require user to flash the screen to find pen
	One-to-one input only (zero order control)
	May be cumbersome to use with alternate, incompatible entry methods, like the keyboard

Figure 7.4 Control selection guide showing advantages and disadvantages of different computer input devices. (*Reproduced from NASA-STD-3000, "Man-Systems Integration Standards," 1987.*)

Human factors information on controls usually shows minimum and maximum size, range of movement, and resistance values. Control guidelines also specify the spacing required between controls. Display guidelines specify lettering size requirements, color requirements, and movement rate. Computers and the video display terminals promote new computer system standards. The United States Human Factors Society and the American National Standards Institute just completed a standard for designing computers and video display terminals (ANSI/HFS100-1988).

Often system designers do not design a specific control or display. They select them from catalogs specializing in selling controls and

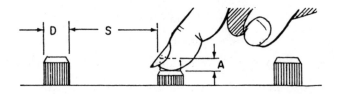

	DIMENSIONS		RESISTANCE		
	DIAMETER D			Different	
	Fingertip	Thumb or Palm	Single Finger	Fingers	Thumb or Palm
Minimum	9.5 mm (3/8 in.)	19 mm (3/4 in.)	2.8 N (10 oz.)	1.4 N (5 oz.)	2.8 N (10 oz.)
Maximum	25 mm (1 in.)		11 N (40 oz.)	5.6 N (20 oz.)	23 N (80 oz.)

	DISPLACEMENT	
	A	
	Fingertip	Thumb or Palm
Minimum	2 mm (5/64 in.)	3 mm (1/8 in.)
Maximum	6 mm (1/4 in.)	38 mm (1-1/2 in.)

	SEPARATION S			
	Single Finger	Single Finger Sequential	Different Fingers	Thumb or Palm
Minimum	13 mm (1/2 in.)	6 mm (1/4 in.)	6 mm (1/4 in.)	25 mm (1 in.)
Preferred	50 mm (2 in.)	13 mm (1/2 in.)	13 mm (1/2 in.)	150 mm (6 in.)

Note: Above data for barehand application. For gloved hand operation, minima should be suitably adjusted.

Figure 7.5 Control design guide showing push-button design criteria. (*Reproduced from MIL-STD-1472D, "Human Engineering Design Criteria for Military Systems, Equipment and Facilities," Department of Defense, March 1989.*)

displays. Design guidelines provided by the human factors profession helps designers select the proper controls and displays from the catalog.

7.2.5 Human size and capabilities data

Human factors supplies designers with information on human sizes and capabilities. This information falls into the following categories:

1. Anthropometric data on human size and mass
2. Physical strength data

3. Human movement and reaction times

4. Mental processing and memory capabilities

5. Perceptual capabilities (vision, hearing, touch, kinesthetics, etc.)

Human size and capability data are in tables and charts. Because humans vary, the tables and charts give a range of measurements. These values are called percentiles. If a person is 5th percentile in height, for instance, this means that 5 percent of the population is shorter and 95 percent of the population is taller. You might be tempted to use the average or 50th percentile value. This will not work in a design. You will probably make most people unhappy if you design for the average person. Very few, if any, people are "average." Your design will be too big for the smallest person and too small for the largest person.

To apply human size and capability data you first determine who is going to use your equipment. Surveys show occupational groups differ in size and capabilities. Anthropometric data and human capability data exist for astronauts, military personnel, law enforcement personnel, children, and many more groups. Use the data for the group that most closely approximates the people you are designing for.

Next, select the range to accommodate. Most military and government contracts require a design to accommodate 90 percent of the population. Designers should design for the largest number of users. Consider the "worst case" person when designing (see Figure 7.6). Once the design fits the population extremes, make sure that everyone else in the population can use the design. The hardware may have to adjust. Automobile seats and steering wheels adjust to accommodate a large range of adults.

The following examples illustrate how to use the human size and capability data. If you are locating an overhead control for a space shuttle pilot, the NASA document NASA-STD-3000, gives data for the astronaut population that will use the equipment. The standard shows that the overhead reach of a 5th percentile person operating controls is 42 in above the seat pan. The standard also shows that 95th percentile operator must have 39 in above the seat pan for head clearance. By using information in the NASA standard you determine that a 95th percentile operator may bump a control placed within the reach of a 5th percentile operator. You have a choice: Put the control somewhere else or put the control above the seat and have the seat adjust up and down. If you think 6 in will be enough clearance for a 95th percentile operator to get in and out of

TASK	"WORST CASE" PERSON DOING TASK
Walk through a narrow passageway	Large
Lift a box	Weak and frail
Reach overhead	Small stature and short arms
Reach under seat for adjustment while seated	Long torso, short arms
Boring or dangerous task	Unmotivated
Respond to an indicator light	Poor eyesight, color blind, distracted

Figure 7.6   Designs should accommodate the "worst case" person.

the seat, then design the seat so that the largest individual can lower it 3 in. Remember, however, adjustable seats influence control and display locations.

Designing an automobile hand brake provides another example. You must decide how much force a person will need to engage the brakes. Human factors data show that the weakest person cannot pull over 16 lb with their right arm (depending on the elbow angle). The hand brake should therefore engage at 16 lb or less. However, a strong person (95th percentile strength) can pull nearly 190 lb. You should therefore design the hand brake to engage with 16 lb of pulling force and to withstand at least 190 lb before failing.

Test subject selection is another application for human size and capability data. For example, assume you want to determine if people can apply sufficient force on a brake pedal to stop a car. Find human factors data on leg strength and have your potential subjects repeat the leg strength test. Select the subjects that come closest to the weakest test score. Use these subjects to test your foot pedal design.

7.2.6   Environmental design guides

The human factors profession has data defining environmental limits for human comfort and safety. The list below briefly describes environmental design guides in human factors literature:

1. *Temperature, humidity, and ventilation:* Physical comfort depends on a combination of these factors. Human factors has comfort envelopes for temperature, humidity, and ventilation. Humans will do tasks poorly outside these envelopes. Too far outside the comfort envelope and the human will die.

2. *Light:* Tables give minimum and recommended illumination levels for specific work areas or tasks. Figure 7.7 is a typical illumination requirements table.

3. *Noise:* Tables show maximum noise levels. Too much noise can damage hearing. There are worldwide standards to limit noise levels and exposure time. Other noise standards give maximum

| | ILLUMINATION LEVELS | |
| | LUX* (FT-C) | |
WORK AREA OR TYPE OF TASK	RECOMMENDED	MINIMUM
Repair work:		
general	540 (50)	325 (30)
instrument	2155 (200)	1075 (100)
Scales	540 (50)	325 (30)
Screw fastening	540 (50)	325 (30)
Service areas, general	215 (20)	110 (10)
Stairways	215 (20)	110 (10)
Storage:		
inactive or dead	55 (5)	30 (3)
general warehouse	110 (10)	55 (5)
live, rough or bulk	110 (10)	55 (5)
live, medium	325 (30)	215 (20)
live, fine	540 (50)	325 (30)
Switchboards	540 (50)	325 (30)
Tanks, containers	215 (20)	110 (10)
Testing:		
rough	540 (50)	325 (30)
fine	1075 (100)	540 (50)
extra fine	2155 (200)	1075 (100)
Transcribing and tabulation	1075 (100)	540 (50)

Note: Some unusual inspection tasks may require up to 10,000 lux (1,000 ft-C)

Note: As a guide in determining illumination requirements the use of a steel scale with 1/64 inch divisions requires 1950 lux (180 ft-C) of light for optimum visibilty.

*As measured at the task object or 760 mm (30 in.) above the floor.

Figure 7.7   Environmental design guide showing illumination requirements for specific work areas and tasks. (*Reproduced from MIL-STD-1472D, "Human Engineering Design Criteria for Military Systems, Equipment and Facilities," Department of Defense, March 1989.*)

noise levels for sleeping, effective communications, and productivity.

4. *Motion:*   Ships, land vehicles, and aircraft vibrate and accelerate. Standards exist to define motion acceleration and frequency limits for human safety, comfort, and proficiency.

5. *Toxic hazards:*   Standards define ingestion, respiration, and direct contact exposure limits for toxic substances. Standards also define exposure limits to ionizing and nonionizing radiation.

7.3   Measurement Tools

The human factors profession uses tools to measure the working environment, human size, and human performance. The table in Fig. 7.8 summarizes the human factors measurement tools. The following sections describe the tools.

TOOL	DESCRIPTION	APPLICATION
Human Size Measurement Tools	Devices to measure and record body length, body mass, and range of joint movement	Data used to design hardware to fit the human
Human Performance Measurement Tools	Tools to measure mental and physical performance limits of humans. Tools include: • Timer • Video and Audio Recording Equipment • Cardiovascular Measurement Tools • Respiration Measurement Tools • Strength Measurement Tools • Electrical Sensors of Muscular Activity	Data used to evaluate jobs, select workers, determine job procedures and schedule, and design workstations
Environment Measurement Tools	Instruments to measure and record environmental factors that may effect human health and performance: • Sound Level Meter • Light Meter • Thermometer • Accelerometer • Anemometer	Data used to evaluate working environments and design equipment to control the environment

Figure 7.8   Human factors measurement tools.

7.3.1 Human size measurement tools

Human body lengths are measured with tapes and calipers. The measurements are usually between standardized points on the body called "landmarks." These measurements allow body sizes comparisons. Human body mass is measured with scales. The body volume and mass distribution is measured by placing the person in water and measuring the displaced water. Body lengths, volume, and body mass distribution can be measured using multiple cameras and a computer to combine the data points. This method, stereophotometrics, is becoming more popular because of the speed of data collection and analysis.

7.3.2 Human performance measurement tools

The human factors profession measures human performance to evaluate designs and to learn more about human capabilities and limits. Figure 7.9 lists tools used to measure human performance and when to apply these tools.

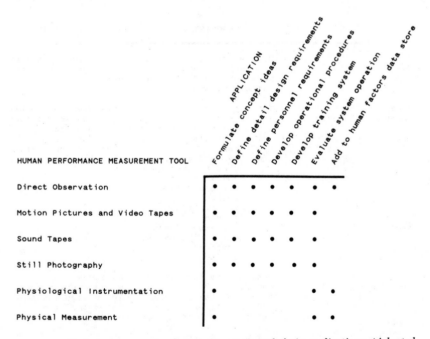

HUMAN PERFORMANCE MEASUREMENT TOOL	Formulate concept ideas	Define detail design requirements	Define personnel requirements	Develop operational procedures	Develop training system	Evaluate system operation	Add to human factors data store
Direct Observation	•	•	•	•	•	•	•
Motion Pictures and Video Tapes	•	•	•	•	•	•	
Sound Tapes	•	•	•	•	•		•
Still Photography	•	•	•	•	•	•	
Physiological Instrumentation	•					•	•
Physical Measurement	•					•	•

Figure 7.9 Tools to measure human performance and their applications. (*Adapted from DOD-HDBK-763, "Human Engineering Procedures Guide," Department of Defense, February 1987.*)

1. *Direct observation:* Direct observation is the most common evaluation method. The person doing the observation should be familiar with the task and equipment. The observer can combine direct observation with additional tools such as a clock, checklists, questionnaires, and interview sheets. We give subjects a list of tasks and make observations about their performance on an audio tape recorder. A tape recorder also acts as a timer. We record when each task starts and ends. When we play the tape back we watch a clock and mark down the times.

2. *Motion pictures and video tapes:* Motion pictures and video tapes are excellent tools for measuring human performance. They provide a permanent event record. Actions that might have been missed by direct observation can be reviewed and analyzed at a later time. The person operating the camera should be familiar with the task and have a list of what to film. Areas such as a large control room may require a wide-angle lens to record what all personnel are doing. Other tasks may require a zoom lens to focus in close on details.

3. *Still photography:* Still photography is a tool to record and communicate human performance study results. Photographs are usually "staged." The analyst observes an event and then has a subject repeat the action in front of a camera. In analyzing hardware we have used two test subjects, one 95th percentile and 5th percentile. We have each subject operate the system and note problems. If we see a problem we want to record and communicate to others (designers and engineers), we have one or both subjects repeat the task and photograph the problem. Instant photography is sufficient for informal communications. Usually, however, the time and expense of the study warrants a more permanent record on 35 mm or larger format film.

4. *Sound tapes:* A communications record is often an inexpensive way to understand human performance. The aviation industry uses the "black box" aboard aircraft to keep a constant communications record. Sound tapes can be used alone or in combination with other tools such as direct observation.

5. *Physiological instrumentation:* Physiological instrumentation can measure the affects of a task on the human. Human performance information helps determine human limits. For example, human factors research may want to reduce warehouse worker fatigue. The researchers may measure muscle activity, heart

rate, and respiration of workers lifting and carrying boxes. The physiological data may show that a different work/rest cycle or different box size can reduce demands on the body. Physiological instrumentation can also measure mental and emotional task demands. Physiological instrumentation can monitor:

a. Heart rate
b. Blood pressure
c. Perspiration
d. Brain electrical activity
e. Body temperature
f. Muscle electrical activity
g. Respiration rate and volume

Physiological instrumentation can be costly and time consuming. The equipment is expensive and test setup and data analysis require medical expertise. The test subjects must be very committed to the study. Physiological instrumentation is therefore not commonly used during system design and development. Human factors researchers use physiological instrumentation to set the human performance boundaries and establish human factors design standards.

6. *Physical measurements:*   The human factors profession uses tools to measure muscular strength and body movement. In designing equipment that requires muscular strength, it is simple to measure or calculate the force required and then consult a human factors strength table. If a design exceeds human limits, then redesign. Often, however, data do not exist and you must measure human strength. There are two types of strength: static strength and dynamic strength. The table in Fig. 7.10 lists tools used to measure each strength type and comments on how to use each tool.

Videocameras record and measure speed and range of body movement. Analysts place lights or reflective tape on the subject's body and define a static reference grid. The grid can be in back of the subject, on a screen in front of the subject, on a videoscreen, or in computer software. The analyst tapes the subject doing a task. Motion speed and range are determined by stopping the tape at intervals and noting the body landmark position in the reference grid. There are computer systems that analyze tapes and give a summary of the range, speed, and frequency of movement.

The oculometer is a useful tool for designing displays to min-

MUSCULAR ACTIVITY	EXAMPLE TASKS	TOOL TO MEASURE STRENGTH	COMMENT
Static Strength (little or no movement)	Pull auto hood release handle, hold a box, push on a brake pedal	Push, or pull torsional spring gages	The push/pull measurement limits should be from 1/4 ounce to 250 pounds. The torque measurement should be from 1/2 inch-pound to 160 foot-pounds.
			Have the subject hold force for at least 4 seconds. Measure the force at 1 second and 4 seconds after application. Average the forces.
Dynamic Strength	Turn a steering wheel, lift and carry boxes, push or pull a cart	Dynamic Tests	Weight can be added or subtracted from moved object to get maximum weight. Difficult to standardize the findings.
		Psycho-physical Tests	Test subject evaluates effort, fatigue, and comfort and adjusts weights and level of exertion. Difficult to standardize the findings.
		Isokinetic Strength Tests	Test subject applies force using a pulley device that keeps speed of movement constant regardless of force. Test results are simpler to standardize.

Figure 7.10   Tools to measure human strength.

imize visual search time. The oculometer tracks the path of the eye across a surface. The designer can experiment with different display arrangements or visual signaling methods.

7.3.3   Environment measurement tools

Human productivity, comfort, and safety depend on the working environment. There are guidelines to define comfortable and safe environments. The human factors profession uses tools to measure the environment and compare these findings with the guidelines. Tools can be used to evaluate an existing workplace or to test and evalu-

TOOL	PURPOSE	MEASUREMENT RANGE (APPROXIMATE)
Photometer	Measures ambient illumination on task. Used to evaluate ability to see task.	.005 to 25,00 foot-candles
Spot Brightness Meter	Measures small area of brightness coming from a display. Used to evaluate display legibility	.01 to 25 foot-lamberts
Sound Level Meter and Analyzer	Measures noise levels. Used to evaluate hearing damage risk, ability to concentrate, and annoyance.	10 to 150 dB Gives sound level for standard weighted noise curves (A and C) and provides octave band analysis for frequencies from 63 to 8000 Hertz.
Vibration Meter and Analyzer	Measures amplitude and frequency of workplace movement. Used to evaluate effects on human comfort, performance, and safety.	Acceleration from .01 to 25 Gs. The analyzer determines amplitudes at selectable frequency bands from 2.5 Hz to 25 kHz.
Thermometer	Measures air, surface, and liquid temperatures. Used to evaluate human exposure to cold or heat stress and to burns or freezing due to surface contact.	32 to 200 degrees Fahrenheit
Anemometer	Measures air flow. Used to evaluate chilling or ventilation of personnel.	0 to 1000 feet per minute
Hygrometer or Psychrometer	Measures relative humidity by wet and dry bulb method. Used to evaluate human comfort.	0 to 100 percent relative humidity
Gas Tester	Measures toxic gas concentration levels. Used to evaluate personnel safety.	Depends on type of gas

Figure 7.11   Tools to measure the environment. (*Adapted from DOD-HDBK-763,* "*Human Engineering Procedures Guide," Department of Defense, February 1987.*)

ate preliminary designs. The table in Fig. 7.11 lists a set of environment measurement tools. The table is adapted from a Department of Defense handbook (DOD-HDBK-763).

7.4   Analytical Tools

Analytical tools developed by the human factors profession help to understand the human in the work environment. Information gathered from using these tools helps produce better human engineered products and expands human factors design guidelines. Human factors analysis techniques and methods, which is a type of human factors tool, were discussed Chap. 6. This chapter discusses simulators, mockups and models, prototype software, and research tools. Figure 7.12 summarizes human factors analytical tools.

7.4.1   Simulators

Simulators are probably the most expensive human factors analysis tools, but they have many uses. The simulator puts the human in a realistic work environment and allows the analyst to test and observe performance. NASA, the Department of Defense, and major airline companies have spent millions of dollars on the design and maintenance of simulators. Simulators have the following uses:

1. *Procedures modification:* NASA uses simulators to evaluate different methods for astronauts to deploy satellites or capture them for repair.

2. *Hardware modification:* Flight simulators can modify aircraft flight characteristics and evaluate pilot performance without "cutting metal."

3. *Test human limits:* Simulators can test human performance limits without risking catastrophic failure.

4. Operator training.

5. Operator evaluation.

NASA and the Department of Defense have several simulator facilities around the country. When planning a human factors effort on a design project consider using simulators to resolve critical human factors issues. Investigate the cost of simulators. A simulator may be the least expensive way to arrive at a correct answer.

7.4.2   Mockups and models

Mockups and models differ from simulators because they do not represent the system procedures. A system consists of people, hard-

ANALYTICAL TOOL	DESCRIPTION	APPLICATION
SIMULATOR	A system that reproduces real system job demands and allows human factors professionals to measure and record human performance	Data from simulators are used to evaluate and modify system design concepts before they are finalized. Simulators are also used as a less expensive and safe way to train people to operate existing equipment (such as a flight simulator)
MOCKUP, MODEL, AND MANIKIN	A two or three dimensional physical simulation. A manikin is a model of the human body.	Mockups and models are used to evaluate human interface with designs before the designs are complete. Test subjects can evaluate mockups and scaled manikins can be used to evaluate models and engineering drawings. Mockups and models are also used as training aids after the system is built.
COMPUTER ANALYTICAL AID	Software that stores and processes information about the human role in system operation and maintenance	Speeds human factors analysis. Provides format for human factors data storage and retrieval.
RAPID PROTOTYPE SOFTWARE	Computer software that simulates the interface between the computer and the computer user	A non-programmer can simulate a computer program for human factors evaluation before software code is finalized.
COMPUTER BIOMECHANICAL MODEL	Three dimensional human model in computer software. Analysts can position the scaled models in computer aided engineering drawings.	Output used to predict and analyse human physical interface with the design (reach limits, clearances, visibility, and strength limits).
COMPUTER DATABASES	Databases containing human factors information, standards, and design guidelines	Rapid data access through key word search. Tele-communications permit information research and retrieval without leaving the work site.

Figure 7.12   Human factors analytical tools.

ware, and software. The simulator incorporates all three. Mockups and models are not interactive, you cannot "plug them in" and evaluate system procedures. Designers and human factors professionals use mockups and models to evaluate a design's physical characteristics:

1. Reach to hand and foot controls

2. Visibility of displays, outside workstation, and other personnel

3. Room to move inside the workstation

4. Equipment arrangement and accessibility

Mockups and models can be either very inexpensive or very expensive. We have "mocked up" a control panel in the early design stages using Post-it notes with controls and displays sketched on them. We arranged the notes to get the most effective panel layout for eye and hand movement. In the early stages of a design, start with the least expensive mockup. The mockup should be as flexible as the design. Cardboard sheets filled with a plastic foam core are inexpensive and excellent mockup construction materials. Foam core board mockups do not require machinists or carpenters to construct. They can be cut with a knife and glued with a hot glue gun. The sheets are strong and you can use them to build the walls and ceiling of an entire control room. Engineering control and display drawings can be taped onto the foam core sheets.

You may need a more expensive mockup. While working for a company designing military vehicles we found the driver station in a military vehicle to be a difficult human factors problem. The driver must be able to operate the vehicle under armor protection and look through periscopes to see. Although at times unsafe, the drivers have better visibility if they sit up high with their head outside the vehicle hatch. To operate the vehicle from both positions the seat must move up and down 12 in. The seat movement usually requires adjustability in pedals, the gear selector, and the steering controls. A driver station this complex is difficult to evaluate from a drawing. A mockup was required. The mockups had periscopes, adjustable seats, and adjustable controls.

A complex mockup requires skilled workers to build. I noticed one of the mockup carpenters in the company I worked for was calling me "bump and grind." I tried to ignore the comment, thinking maybe I should avoid him. I finally asked. He said, "I build a mockup. Then you bump yourself on a sharp corner and tell me to grind it off."

A more expensive mockup has other uses besides human factors. Mockups can be used by engineers to design routings for cables and conduits, design storage areas, and to check for interferences between components. A realistic mockup can communicate design ideas and progress to the customer. At the end of a program the mockup can be a training device.

The human factors profession occasionally uses models. We own an artist manikin close to one-sixth scale. We once used foam core to build a scale model of a passageway we were concerned about. Using the manikin and the model, we showed the customer how it would be nearly impossible to use the passageway as designed.

A useful tool for evaluating engineering drawings and sketches is a two-dimensional, articulated, clear plastic manikin. Figure 7.13 shows an example. Manikins usually show a side view and are available in scales commonly used for engineering drawings ($\frac{1}{10}$, $\frac{1}{8}$, $\frac{1}{4}$, $\frac{1}{2}$). The manikins represent human sizes. They are sold commercially and there is a template for making them in the military document MIL-HDBK-759A. You use the manikins by laying them over a scaled drawing in positions you expect operators will be in. You can evaluate a design in the early stages by estimating clearances, reach distances, and visibility (by knowing where the eyes will be).

7.4.3 Computer

The computer is a valuable tool for human factors. The following list describes jobs a computer can help with in forming a design.

1. *Analytical aid:* Computer programs are available to help with task analysis, functional analysis, and training program development. Chapter 6 discusses human factors methods in more detail.

2. *Rapid prototyping:* Computer tools are available to simulate or "mockup" computer programs for human factors analysis. These tools are called rapid prototyping software. Rapid prototyping software allows a nonprogrammer to mockup the part of the program that the operator will use directly. Analysts can use rapid prototyping software to design menus, select colors, locate windows, and select function keys. The software programmer can then incorporate the rapid prototyping results into the program.

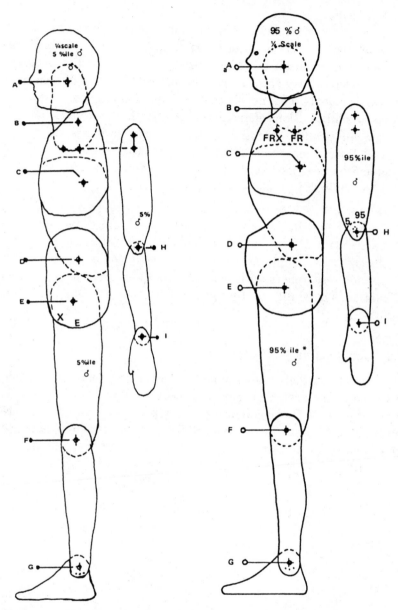

Figure 7.13 Two-dimensional manikin. (*Reproduced from MIL-HDBK-759A, "Human Engineering Guide for Army Material," Department of Defense, 1981.*)

3. *Biomechanical models:* Human biomechanical models can be incorporated into computer-aided design. The three-dimensional human models can be scaled, sized, and positioned to evaluate any design. They show more information than two-dimensional plastic models. The models show reach distances, working room, and visibility. Biomechanical models even have strength data, so you can estimate strength for specific body positions. Computer models are sometimes difficult to use. To get the best use of biomechanical models, you should be proficient with the software and not dependent on a programmer.

4. *Database research:* Many human factors standards and guidelines are now in computer databases. You can use key words to access data. Data and research findings are also available through telecommunications. Worldwide computer networks provide researchers with instant access to human factors information. The ability to use these databases is a valuable human factors tool.

7.5 Conclusions

Human factors tools allow human factors professionals or anyone involved in the design to improve products or systems. These tools make it possible for people in the design process to analyze and apply validated human factors information to their designs. Chapter 8 gives additional information on a very important tool: the design standard.

Chapter

8

What Are Design Standards And How Do I Use Them?

8.1 The Birth of a Human Factors Standard

In this chapter we will explain human factors design standards and how to use them. Human factors design standards protect humans, assure work productivity, and document human factors design solutions.

The Department of Defense, NASA, the automotive industry, and the electronics industry all have human factors design standards. Human factors professionals, design professionals, and equipment users create human factors design standards based on human factors tests, research, and experience.

The following is a brief discussion of the birth of a standard to help you understand how to apply and improve a human factors standard. We helped author NASA's human factors standard, NASA-STD-3000, Man-Systems Integration Standards (MSIS). This family of documents is the first human factors standard applicable to all United States manned space programs. Because we helped develop this standard we are going to use it as an example throughout this chapter.

NASA awarded Boeing Aerospace Company the prime contract to write the standard. Eleven authors gathered information about humans in space from research and NASA design guidelines. Five world known experts in space and human factors worked with the authors. Boeing invited approximately 50 representatives from prime aerospace contractors, support contractors, NASA centers, and other government agencies to participate in the review process. These 50 people were called the *Government/Industry Advisory Group* (GIAG).

The authors and the experts wrote draft chapters. Every 3 months the authors, the experts, and the GIAG met for one week to review and recommend changes in chapter organization and content. The authors reviewed the comments, conducted research, and revised the chapters during the interim between the meetings.

Our goal was to keep the material accurate and usable. We included information in the standard if the authors, experts, and GIAG members felt it was valid. We reviewed and evaluated all research studies. If the research was poorly done or not well documented, we did not include the information in the standard. Although tempting, we did not have the time or budget to repeat research studies or initiate new ones.

The biggest challenge was to determine best level of detail. If a standard is too general it will not provide enough guidance for designers. Too many requirements could reduce design creativity and eliminate an effective and less expensive design solution. In a zero gravity space environment, water showers have not been successful. If you are designing a personal hygiene station in a space vehicle and the standard specifies a person shall bathe with water, the standard could eliminate the possibility of another form of bathing. A compromise might be: Provide means to clean the entire body.

8.2   How Do I Use a Human Factors Standard?

You use human factors standards at all design stages. Figure 8.1 summarizes ways to use human factors design standards throughout a design process. We will explain Fig. 8.1 and give examples using actual standards.

If you are responsible for meeting a human factors standard, start by reviewing the standard. Become familiar with the topics covered by the standard. Don't memorize the standard; you only need to know what areas the standard covers.

First use the standard to allocate functions. Most human factors

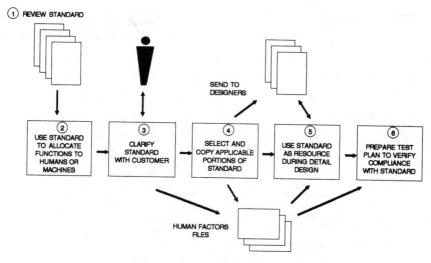

Figure 8.1   How to use a human factors standard.

standards give human capability limits. This information will help decide if a human or machine should do a function. For example, you have a military contract requiring loading ammunition from the ground to the back of a 3-ft-high truck bed. Your design will have to meet the Department of Defense standard MIL-STD-1472 and this standard contains human lifting limits. If the ammunition is too heavy to meet the standard then you will have to allocate the function to a mechanical system such as a pulley or hydraulic lift.

Once you know the equipment in the system, find sections in the standard that apply to the equipment. Copy these standard sections. Keep a copy for your files and distribute copies to appropriate designers. Clarify all questions about the standard with your customer (the agency you are designing the product for).

Assume that you are responsible for human factors in the sleeping compartment design for the NASA space station. Your contract specifies NASA-STD-3000, MSIS, vol. 1. We will use this example to show how to use a human factors standard.

First look in the standard to find sections about sleeping compartments. The MSIS has a keyword list. When you look up "sleeping" you will find reference to Sec. 10.4, "Crew Quarters." Figure 8.2(a), (b), and (c) shows Sec. 10.4, pages 10.4-1, 10.4-2, and 10.4-3, respectively. When you review Sec. 10.4 you will discover that the standard separates the considerations and the requirements into

---

NASA-STD-3000

**INDIVIDUAL CREW QUARTERS
DESIGN CONSIDERATIONS**

10.4 CREW QUARTERS
{O }

10.4.1   Introduction
{O }
This section covers design consid-erations and requirements for the design and layout of private activity and sleeping quarters for an individual crewmember in a microgravity environment.

(Refer to Paragraph 8.6, Envelope Geometry For Crew Functions, for volume envelope design considerations and requirements.)

(Refer to Paragraph 8.3, Crew Station Adjacencies, for information on crew quarters location considerations and requirements.)

(Refer to Paragraph 7.2.4, Sleep, for information on sleep and its relationship to health.)

**10.4.2   Individual Crew Quarters
Design Considerations**
{O }
The following design considerations apply to the design and layout of crew quarters.

a.   Mission Duration and Privacy - The amount of volume required for each crewmember is dependent on the duration of the mission. As the mission becomes longer the need for privacy increases. There are several design solutions for individual privacy. One of these solutions is described in this section: private quarters for individual recreation and sleeping. Other arrangements for privacy include:

1.   Dormitory sleeping and private areas available to each crew-member.

2.   Shared private quarters so that two crewmembers on different shifts share the same quarters.

3.   Quarters for two individuals who want privacy (i.e., married couples).

4.   Expanded function quarters which might include full body wash facility, waste management facility, office, private dining, or meeting facility.

b.   Functional Considerations - The design and layout of the crew quarters depends on the functions that are to be performed. Figure 10.4.2-1 shows the functions that might occur in individual crew quarters and the design considerations to accommodate these functions.

**10.4.3   Individual Crew Quarters
Design Requirements**
{O }
The following are design requirements for one-person individual crew quarters:

a.   Communications - Two way audio/visual/data communications system shall be provided between the crew quarters, other module areas, and the ground. The system shall have the capability of alerting the crew quarters occupant in an emergency.

b.   Environmental Controls - Independent lighting, ventilation and temperature control shall be provided in crew quarters and shall be adjustable from a sleep restraint.

(Refer to Paragraph 5.8.3, Thermal Design Requirements, for thermal and ventilation requirements, and Paragraph 8.13.3, Lighting Design Requirements, for lighting requirements.)

c.   Noise - The noise levels in the crew quarters shall be as low as possible during sleep periods.

(Refer to Paragraph 5.4.3.2.3.1, Wide-Band Long-Term Noise Exposure Requirements, for permissible noise levels.)

d.   Movement - The vibration and acceler-

10.4-1

Figure 8.2(a)   Example human factors standard. (*Reproduced from NASA-STD-3000, "Man-System Integration Standards," March 1987.*)

different sections and different type fonts. The design consider-ations in Sec. 10.4.2(a) is only background information. The design must meet the design requirements under Sec. 10.4.3 (a) and (c). You should copy entirely Sec. 10.4.3 and take these requirements to the designers responsible for the crew quarters.

INDIVIDUAL CREW QUARTERS
DESIGN CONSIDERATIONS

NASA-STD-3000

Function	Design considerations	Reference paragraphs
Wake up	Alarm or annunciator	9.4.4
Dress	Adequate volume	8.6.2.3
	Privacy	8.6.2.4
	Restraints	11.7
	Clothes and personal items storage	10.4.3
Straighten/clean quarters	Bedding storage	
	Vacuum and wipe capability	13.2
Groom	Adequate lighting	10.2.3.5
	Mirrors	10.2.3.5
	Stowage for grooming supplies	10.2.3.5
	Proximity to personal hygiene facility	8.3.2.2
Exit	Lock for personal items	
	Properly configured door and path	8.8, 8.10
Enter	Properly configured door	8.8, 8.10
Relax	Communications with friends or family at home	10.7.2
	Entertainment materials: books, audio and video entertainment, games, etc.	
	Adjustable lighting	8.13
	Window	8.11, 11.11
	Ventilation and temperature control	5.8
	Restraint	8.9, 11.7, 11.8
	Snack storage and cleanup capability	
	Aesthetically pleasing environment	8.12
Prepare for sleep	Clothing and bedding storage	10.10.3
	Proximity to personal hygiene and body waste management facility	8.3
	Privacy	
Sleep	Quiet	5.4
	Privacy	
	Adjustable lighting	8.13
	Bedding	
	Restraints	8.9, 11.7, 11.8
	Ventilation and temperature control	
	Stability (minimum vibration and acceleration)	5.5
Emergency	Alarm	9.4
	Two way communications with other crewmembers or ground control	9.4.3
	Emergency lighting	8.13
	Properly configured door and path	8.8, 8.10
Work	Privacy	
	Workstation	

Figure 10.4.2-1. Individual Crew Quarter Functional and Design Considerations

MSIS-347

10.4-2

Figure 8.2(b)   (Continued)

The designers will need more detailed requirements from the standard. To find more details for this example look at the design requirement in Sec. 10.4.3 (b). It states that the individual crew quarters should have independent lighting, ventilation, and temperature controls. Designers need to know the acceptable lighting, heating, and ventilation range. They also need human factors standards for control design and location. Notice the design requirements refer you to other standard sections for details on each requirement. Paragraph 8.13 in MSIS lists illumination requirements. When you scan through Par. 8.13 in MSIS you will find two

NASA-STD-3000	INDIVIDUAL CREW QUARTERS DESIGN REQUIREMENTS

ation of the crew quarters shall be minimized to the maximum extent possible.

(Refer to Paragraph 5.5.3.3.3, Reduced Comfort Boundary, for sleep area vibration limits.)

e.   Stowage - Facilities shall be provided in the crew compartment for stowing the following items:

    1.   Bedding.

    2.   Clothing.

    3.   Personal Items.

f.   Compartment Size - For long duration space missions, dedicated, private crew quarters shall be provided for each crewmember with sufficient volume to meet the following functional and performance requirements:

    1.   1.50 m3 (53 ft3) for sleeping.

    2.   0.63 m3 (22 ft3) for stowage of operational and personal equipment.

    3.   Additional free volume, as necessary, for using a desk, computer/-communication system, trash stowage, personal grooming, convalescence, off-

duty activities, and access to stowage or equipment without interference to or from permanently mounted or temporarily stowed hardware. The internal dimensions of the crew quarters shall be sufficient to accommodate the largest body size crew-member under consideration.

g.   Exit and Entry - The opening shall be sufficiently large to allow contingency entry by an EVA suited crewmember.

(Refer to Paragraph 8.10.3, Hatch and Door Design Requirements, for requirements on doors.)

(Refer to Paragraph 14.5.3.5, EVA Passageway Design Requirements, for minimum opening for EVA suited crewmember.)

h.   Privacy - The individual crew quarters shall provide audio and visual privacy for the occupant.

i.   Restraints - Restraints shall be provided as necessary for activities such as sleeping, dressing, recreation, and cleaning.

(Refer to Paragraph 11.7.2.3, Personnel Restraints Design Requirements, for requirements on restraints.)

10.4-3

Figure 8-2(c)   (Continued)

sections that apply directly to the crew quarters: a minimum light level table and Par. 8.13.3.1.3, "Illumination Levels of Sleeping Areas—Design Requirements" (Fig. 8.3).

The information in Fig. 8.3 is confusing. The chart says that the minimum light level for sleeping is 54 lux and the paragraph states that the lights should go off. You may want to assume that the lights should go from off to 54 lux. Document those assumptions in your records. If you feel you do not have enough information, call someone at NASA responsible for the contract and discuss your dilemma. They will appreciate your call and respond with their needs. Have the customer (NASA) answer your questions in writing and keep a copy for you files. Also, you need to know the maximum lighting level. As you can see in the table, the lighting level should be at least 323 lux if the crewmember will be reading in the quarters. Again, check with the customer (NASA) and see what they

8.13.3.1.3 Illumination Levels of Sleeping Areas Design Requirements
(A)

The following requirements apply to the illumination of sleeping areas:

a. The lighting level shall be adjustable from "off" to the maximum for sleeping areas.

b. Minimum lighting of 30 Lux (3 ft. cd), or other means of visual orientation shall be provided to permit emergency egress from sleeping areas.

Area or Task	LUX	(Ft.C.)
GENERAL	108	(10)
PASSAGEWAYS	54	(5)
Hatches	108	(10)
Handles	108	(10)
Ladders	108	(10)
STOWAGE AREAS	108	(10)
WARDROOM	215	(20)
Reading	538	(50)
Recreation	323	(30)
GALLEY	215	(20)
Dining	269	(25)
Food Preparation	323	(30)
PERSONAL HYGIENE	108	(10)
Grooming	269	(25)
Waste Management	164	(15)
Shower	269	(25)
CREW QUARTERS	108	(10)
Reading	323	(30)
Sleep	54	(5)
HEALTH MAINTENANCE	215	(20)
First Aid	269	(25)
Surgical	1076	(100)
I.V. Treatment	807	(75)
Exercise	538	(50)
Hyperbaric clinical lab	538	(50)
Imaging televideo	538	(50)
WORKSTATION	323	(30)
Maintenance	269	(25)
Controls	215	(20)
Assembly	323	(30)
Transcribing	538	(50)
Tabulating	538	(50)
Repair	323	(30)
Panels (Positive)	215	(20)
Panels (Negative)	54	(5)
Reading	538	(50)
NIGHT LIGHTING	21	(2)
EMERGENCY LIGHTING	32	(3)

Reference: 351 with updates
NOTE: Levels are measured at the task or 760 mm. (30 in.) above floor.
All levels are minimums.

MSIS 291 REV. A

Figure 8.3 Human factors standards for space vehicle sleeping-area illumination. (*Reproduced from NASA-STD-3000A, "Man-System Integration Standards," October 1989.*)

want. If reading is not required, the lights can be only 108 lux. Include the information on maximum and minimum lighting levels in the data that you send the designers.

Organize your own file of applicable standard information by equipment components. In our example, we would have a folder labeled "Crew Compartment" or "Sleeping Quarters." You will keep all standards and other materials related to crew compartments in this folder, including:

1. Customer correspondence

2. Correspondence with the designers

3. Design review notes

4. Test plans

5. Task analysis work sheets

You will continue to use the human factors standard when the program moves to the detail design and development stage. Continuing with our space station sleeping quarters example, you analyze tasks involved in getting ready for sleep. You build a mockup of the sleep compartment and discover one of the test subjects has difficulty reaching the light switch from the sleep restraint. Item (b) in Par. 10.4.3 (Fig. 8.2) states that people should be able to reach the light controls from the sleep restraints. (A sleeping restraint is a sleeping bag attached to the wall to keep the sleeper from floating all over the space ship.)

Chapter 3 in NASA-STD-3000 has information on the reach limits of people. The standard shows reach limits in nearly all directions and for both large and small astronauts. Figure 8.4 is from NASA-STD-3000. It shows reach limits in the plane on the body centerline. Thirty more figures in the standard show reach limits in all directions. Use the NASA standard to determine the smallest astronaut's reach limits in the direction of the light switch. Maybe you will have to relocate the switch so everyone can reach it. You can use the standard as a human factors information resource throughout the detail design and development stage.

Finally, use the standard to check your design in the final test stage. Your design must meet the standards. In our example you may decide to test the illumination level in the crew quarters. When planning final human factors tests, consult your standards file for crew quarters. The illumination level should be at least 323 lux (assuming the customer wants reading in the crew quarters). Include this level as the minimum acceptable level in the test plan. Standards also sometimes tell how to do the testing. Figure 8.3 states that the illumination levels should be measured at the task. This means you should hold the light meter at points where the as-

YZ Plane

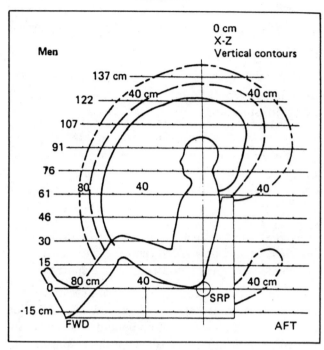

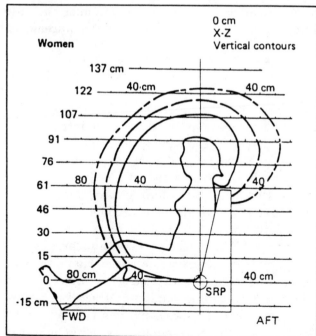

Figure 8.4 Human reach limits. The outer curve is for the ninety-fifth percentile person, the middle for the fiftieth percentile, and the inner for the fifth percentile. (*Reproduced from NASA-STD-3000A, "Man-System Integration Standards," October 1989.*)

tronaut will read the material. Be sure to include this requirement in the test plan.

8.3 What Happens if You Don't Follow Human Factors Design Standards

Failure to follow human factors design standards can have negative consequences and the following paragraphs explain those consequences.

Design standards are normally divided into two types of information:

1. *Design requirements:* These are mandatory design requirements. The designer must follow these requirements. Failure to meet these requirements means defaulting on the contract.

2. *Design considerations:* Design considerations are recommendations and supplementary information to help the designer. The designer does not have to follow the recommendations. The standards' authors want the designers to consider the recommendations. Design considerations provide background information to help the user understand the rationale behind the requirements.

Failure to meet human factors design standards can result in conditions that will injure or kill the people that maintain or operate the system. Unsafe products are bad for the company's image and reputation. In addition, costs can be very high for product liability insurance and for damages awarded to the plaintiff for personal injury or loss of property. The legal system can fault the designer with a poor design if the design did not follow human factors design standards. Even if a contract does not specify a particular standard, you can still be liable for failure to meet the standard. Courts can decide that you were negligent and should have followed practices in the standard simply to protect the consumer.

Systems that do not meet human factors design standards can degrade the operators' and maintainers' performance. Human factors design standards have excellent human factors information and you should use them as a resource, even if a contract does not require them. Controls may be too far away for a small person, a computer program can be very difficult to operate, a warning light may not be visible in bright light. These problems will result in a poorly performing system. Poor system performance results in loss of customers. Customers can be the government, such as the Department of Defense or NASA, or a commercial customer such as a family buying a new car.

8.4 What if I Don't Understand, Like, or Agree with a Human Factors Design Standard?

A standard may have mistakes and may contain out-of-date information. When a human factors standard is not accurate, realistic or does not apply to your specific design, you must notify the agency responsible for the standard. Part III lists standards and the responsible agencies. The agency might be able to clarify the standard and solve your problem. The agency might advise you to notify the customer of your problems and see if you can get an agreement (in writing) to relieve you of meeting the requirement. The agency controlling the standard will want to know if there is better information or an error in the standard.

Human factors standards are meant to be living documents and grow with time. Standards should change and improve as we learn more about the subject. Most standards have a blank comment and correction sheet in the back for you to fill out. Figure 8.5 is a correction sheet from MIL-STD-1472. Organizations responsible for standards regularly schedule meetings among users, human factors professionals, designers, and others, to review these comments and update and correct the design standards.

STANDARDIZATION DOCUMENT IMPROVEMENT PROPOSAL
(See Instructions – Reverse Side)

1. DOCUMENT NUMBER	2. DOCUMENT TITLE

3a. NAME OF SUBMITTING ORGANIZATION	4. TYPE OF ORGANIZATION *(Mark one)*
	☐ VENDOR
	☐ USER
b. ADDRESS (Street, City, State, ZIP Code)	☐ MANUFACTURER
	☐ OTHER *(Specify)*:

5. PROBLEM AREAS

a. Paragraph Number and Wording:

b. Recommended Wording:

c. Reason/Rationale for Recommendation:

6. REMARKS

7a. NAME OF SUBMITTER *(Last, First, MI)* – Optional	b. WORK TELEPHONE NUMBER *(Include Area Code)* – Optional
c. MAILING ADDRESS *(Street, City, State, ZIP Code)* – Optional	8. DATE OF SUBMISSION *(YYMMDD)*

DD FORM 82 MAR **1426**   PREVIOUS EDITION IS OBSOLETE.

Figure 8.5 Correction and comment form from a human factors standard. *(Reproduced from MIL-STD-1472D, "Human Engineering Design Criteria for Military Systems, Equipment and Facilities," Department of Defense, March 1989.)*

9

How Can Human Factors Improve an Existing Product?

9.1 Product and System Improvements

All designs can use improvement. Even though a detailed analysis may have been done, problems may arise that need correction or improvement. The problem could be inadequate maintenance or poor operator performance. Often these are human factors or human-machine interface problems although the symptoms do not suggest this.

Our purpose in writing this chapter is to explain how you can improve your existing design. We are going to explain human factors methods for improving a system or any component or equipment item in a system. These instructions for improving a system apply to any product. You can improve chemical plants, hotels, or airplanes by examining the equipment, procedures, and people in the system. You can use the same process to improve the design of any smaller product such as a computer printer or dishwasher.

The block diagram in Fig. 9.1 shows the process for correcting a human factors problem in a system or system component. The following sections discuss Fig. 9.1. Section 9.2 explains human factors problems and possible symptoms. If there is a human factors problem, Sec. 9.3 will help you identify it. Section 9.4 covers ways you can correct the problem. Often designers concentrate on changing hardware to correct human factors problems. Section 9.5 gives details on other ways of correcting human factors problems: Changing

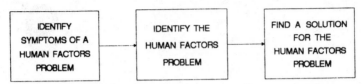

Figure 9.1   Process to correct a human factors problem in an existing system.

the operating procedures and changing the user through selection and training.

9.2   Symptoms of a Human Factors Problem

There are generally two types of problems in an existing system:

1. Problems that cause discomfort, injury, or death to the human

2. Problems resulting from human error or inefficiency

Figure 9.2 lists symptoms for both problems. The symptoms in Fig. 9.2 can be present in an entire system, but also apply to any

HUMAN FACTORS PROBLEM	POSSIBLE SYMPTOMS	WAYS TO INVESTIGATE
Human Injury or Discomfort	High absenteeism	Interview employees or customers
	Worker or customer complaints	Examine employee health
	High rate of worker compensation or personal damage claims	Measure working environment (heat, cold noise, etc.)
	Lack of employee motivation or loyalty	Observe workers or customers
	High employee turnover	Inspect equipment
	Poor product or system sales	
Human Errors or Inefficiency	High system failure rate	Determine how system should operate an assess possibility of human failures
	Poor maintenance	
	Low productivity	Observe workers or customers
	High accident rate	
	Frequent commercial product returns	Inspect equipment

Figure 9.2   Human factors problem symptoms and ways to investigate.

system component. If the symptoms in the table match what is going on in your system then you might have a human factors problem. The last column lists ways to investigate these symptoms to determine if there is a human factors problem.

Any condition where humans are uncomfortable or suffering is a human factors problem. Human factors can help change a system, or system component, to protect people and make them more comfortable. When the user is killed or injured, the problem is apparent. Human factors problems may be more subtle than injury or death. For example, human comfort can be important to the commercial product manufacturers. Customers may buy a product like a suitcase or briefcase simply because the handle is large enough to fit their hand and is comfortable to use. Human factors can help make products and systems more pleasant to use and more appealing.

The other human factors problem listed in Fig. 9.2 is human error and reduced productivity. Human factors can help design a system to optimize human efficiency by improving labels, redesigning control systems, improving training, and designing a better working environment.

Problems other than human factors can cause the symptoms listed in Fig. 9.2. You must investigate problems to determine if they are human factors problems. There are several techniques for determining if there is a human factors problem.

1. *Interview:*   Ask the system users their opinion of the problems. User opinion is particularly important for commercial products. Maintenance tasks such as changing a vacuum cleaner bag, oiling a sewing machine, or changing spark plugs in a car may seem only a minor annoyance. Unfortunately the tasks may influence the owner never to purchase another product from that manufacturer again. However, don't rely on the user's solutions to a human factors problem. The person may give you a personal solution. Also, remember human factors problems can exist without the users telling you about it. Users may not communicate dissatisfaction about a product for several reasons. They may not want to be thought of as complainers, they may be unaware of the problem, or may not trust you and your motives for asking. They may worry about sounding stupid or may not want to admit they purchased a poor or inefficient product.

2. *Observe workers:*   Understand how the system should operate

and look for human errors or risk of human injury. Be inconspicuous; workers may try to hide problems if they know you are observing them.

3. *Inspect equipment:* Equipment users will sometimes temporarily fix a human factors problem, so look closely for these modifications. You can often find temporary modifications on heavy equipment such as a crane, backhoe, grader, or factory machinery. Equipment does not have to remain aesthetically pleasing and users often make changes to improve the human factors. Sharp corners may get padded with foam rubber so the users will not hurt themselves. Short operators may bolt a block of wood to a pedal so they can reach it. Or, users may paint an emergency control red. A control panel may have labels pasted or drawn onto it to act as a reminder for the user.

4. *Measure working environment:* If you suspect the environment is causing human discomfort or injury, measure the environment and compare the data to human factors standards for comfort and safety.

5. *Examine the users' health:* Set up a program to monitor the workers' health before and during the job. Employers do this when they suspect a possibility of job-related hearing damage.

6. *Analyze operations:* Review all manuals and procedures on the system operations and try to identify potential human factors problems. This often works best when combined with the above techniques.

Once you see a possibility of a human factors problem, you can work to identify the problem and solve it. Sometimes the problem will have an immediate and easy solution with minor changes to equipment or procedures. The next section gives steps for defining a human factors problem.

9.3 Defining Human Factors Design Problems

If you determine that you have a human factors problem in your design you must identify the problem and its cause. To find the problem and the cause, look at the tasks the human must do. Tasks that make high demands are true sources of human factors problems. Evaluate each task step in terms of the demands on the human. Look at the steps under all anticipated environmental conditions. If the task is done outdoors, consider darkness or extreme cold or hot conditions. Indoor tasks are usually plagued by high noise levels. Evaluate the following task demands:

1. *Perceptual demands:*   Assess the task demands on the human senses (primarily hearing and vision).

2. *Mental demands:*   Determine demands on memory and reasoning.

3. *Motor demands:*   What motor skills are required to do the task? This includes strength, dexterity, and physical conditioning.

4. *Stress:*   Does the task have mental or physical stresses which will harm the people, make them uncomfortable, or cause them to do the task poorly?

Evaluate your product's tasks on a worksheet form such as that shown in Fig. 9.3. The worksheet will be your notepad when identifying the human factors problems. If there is more than one worker, use separate worksheets for each person. Start the worksheet by listing the tasks in proper sequence in a column on the left. Next to the task list make columns to score the four task demands listed above. A three-point scale (high, medium, and low demands) is usually enough to rate each task step. Use the right-hand column to write remarks. You can write down the reasons for difficult or dangerous tasks. You can also note applicable research or other information.

When trying to find the reason for a human factors problem, think about system components. Figure 9.4 shows that people, equipment (hardware and software), procedures, and the environment are all components of a system. Any single system component can cause a human factors problem by making tasks more difficult or dangerous. Figure 9.5 shows ways system components affect task demands. When analyzing each task step ask yourself the questions in Fig. 9.5.

The best way to analyze tasks is to monitor someone doing the task. You may use the actual worker or a test subject. It is also very helpful to try to do the task yourself.

After you have finished evaluating the task demands you should know what the human factors problems are and their cause. You are now ready to correct the design problems.

| NO. | TASK | TASK DEMANDS* | | | | REMARKS |
		PERCPT	MENTAL	MOTOR	STRESS	

* H = HIGH
M = MEDIUM

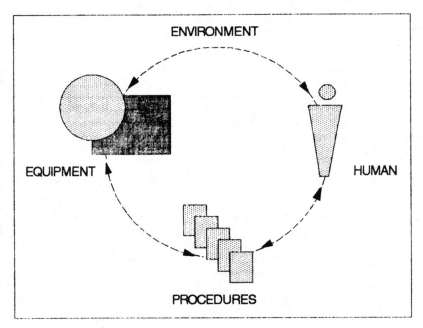

Figure 9.4   Systems have four components: humans, environment, procedures, and equipment. Any systems component can cause a human factors problem.

9.4   Correcting a Problem

After you have identified the problem in a design, you can begin to reduce or eliminate the problem. We will list the steps for correcting a human factors problem and give an example to illustrate the steps.

A company manufactures a hand-held power drill. They have received several complaints and one product liability suit because the drill can inadvertently lock into the "on" position. The drill continues to run even after the operator releases the trigger. Your drill design has a lock on the trigger allowing the user to operate the drill for long periods (buffing an automobile for example) without holding the trigger in. Through the methods described in the previous section, you confirm that the trigger lock switch can be inadvertently engaged by the user. Use the following steps and a product improvement worksheet such as that provided in Fig. 9.6 to help you find a solution:

| | SYSTEM ELEMENTS | | | |
	EQUIPMENT	PROCEDURES	HUMANS	ENVIRONMENT
TASK DEMAND: PERCEPTUAL	Are displays difficult to see? Can warning signals be heard? Can color blind persons use software? Do controls provide adequate tactile feedback?	Can labels be seen and read? Does the job allow the worker to remain at a control station will they will see a warning light? Should the instructions have more illustrations?	Are face to face conversations needed for proper communication?	Can the task be done in all anticipated illumination levels? Will noise levels mask out warning signals or communications? Will glare interfere with tasks? Will gloves reduce tactile feedback from controls?
MENTAL	Do displays provide a complete picture of what is happening to the system? Are software "help" screens adequate? Do labels explain control functions?	Are procedures too complex for some workers? Are instructions and technical manuals easy to understand?	Is training adequate? Are personnel selection procedures compatible with job demands? Are decisions made by properly informed and trained personnel?	Is oxygen level adequate for proper mental functioning? Will heat and humidity degrade mental performance? Will noise effect concentration?
MOTOR	Are forces too high? Can all people reach the equipment? Does the workstation restrict movement of large persons? Are the movements too complex for younger persons? Can physically handicapped persons use the equipment?	Is the work pace and schedule too fatiguing to maintain performance? Do the instructions require physically impossible tasks?	Are personnel selection procedures compatible with job demands? Is training adequate?	Is temperature and humidity too high to maintain work pace? Will gloves prevent operation of switches and push buttons? Are control surfaces too hot or cold to touch? Is lighting adequate for fine motor tasks? Do vibrations & accelerations degrade motor performance?
STRESS	Does equipment have sharp edges or corners? Are seats comfortable for all workers? Are safeguards adequate to prevent worker injury? Are controls and located to prevent awkward working positions?	Are work pace and schedule within worker capabilities? Are safety procedures adequate?	Are job responsibilities clear to the worker? Are workers sufficiently compatible to do the job?	Is the worker exposed to toxic substances? Will noise levels make the job unpleasant or damage hearing? Will vibrations & accelerations make the job unpleasant or injure the worker?

Figure 9.5 Questions to ask when analyzing tasks.

1. *Select the function:*   To correct the human factors problem look at the function associated with the problem. The design is one way to accomplish a function and you must look at alternative designs to accomplish the same function.

Select the function level you are willing to change in the design. Functions consist of several levels and small subfunctions make large-scale functions. In our drill example, the general function associated with the trigger sticking problem is: Operate the drill without holding in the trigger for jobs like polishing a car. You may decide to drop this function from your design or design another system for polishing cars. If you are not able to make a design change at this function level, you must look at subfunctions. You may choose to look at the following subfunction: Engage and disengage trigger lock. Write this function in the left column of the worksheet in Fig. 9.6.

2. *Define present design:*   Under the "Allocated To" column in Fig. 9.6, list the hardware, software, people, and procedures that accomplish the function. Describe how the user engages the switch. Next, in the "Human Factors Problems" column list the problems you found.

3. *Investigate other designs to accomplish the functions:*   Use the last two columns in the product improvement worksheet to record ideas for resolving the problem. You will probably work with system designers to fill in these columns. In the "Rationale" column, list human factors guidelines and other rationale to support the alternative solutions.

4. *Select design concept:*   You will need to test and analyze the alternative solutions. You will work with designers, production personnel, marketing personnel, contract managers, and accountants to select a final design change.

Below are listed changes you can use to eliminate a human factors problem:

Provide the user with protective safety equipment

Remove the user from the hazard

Modify operating or maintenance procedures

Change the environment for the user (insulation, air conditioning, increased illumination, etc.)

HUMAN FACTORS IN PRODUCT IMPROVEMENT

FUNCTION	PRESENT SYSTEM		DESIGN CONCEPTS	
	ALLOCATED TO	HF PROBLEMS	ALLOCATE TO	HF RATIONALE
Engage & Disengage Trigger Lock	Operator pushes button on side of drill near trigger to engage trigger lock. Operator pushes button on opposite side of drill to disengage trigger lock.	Trigger lock button is inadvertently engaged by thumb of right hand person while operating trigger	1. Relocate button away from trigger	Operator will have to make intentional motion to engage trigger lock
			2. Replace button with trigger lock that requires two hands to engage	Operator will have to make intentional motion to engage trigger lock
			3. Put interlock on button so that it must be engaged before drill trigger is pushed	Study shows button is inadvertently pressed only after trigger is pressed. Interlock will require an additional intentional step.

Figure 9.6   Human factors worksheet for improving a design. Example using a hand-held power drill.

Change hardware or software

Change instructions

Change or add user training

Modify or add user selection criteria

Although we used a small product (a drill) for our example, the above process will also work for an entire system such as a chemical plant, cruise ship, or space station.

Design changes do not have to be limited to hardware or software modification. You can change how people use your design by changing the selection, training, labels, or procedures. We will discuss this more in Sec. 9.5.

9.5 When You Can't Change Your Design

Our ideal design would be safe and simple for anyone to operate and maintain without training or instructions. There are times you cannot redesign a product or system because of budget, time, or other equally important reasons. If you find this is your problem, your solution may be to change the way the people use the design. This can be accomplished though selection, training, and labeling.

One way to correct a human factors problem without changing the design is to limit the people who use your product or system. By limiting the user you may prevent injury to the user and damage to your product. The state we live in tests us before we can operate an automobile. Industry requires employees to have specific skills to operate equipment and the military and NASA have physical and mental standards. Even commercial products limit their customers with advertising and labels such as "For ages 12 and up," "Software for computer programmers," and "Warning, do not open unless you are a qualified service technician!"

Restricting the user may eliminate the human factors problem, but it may have disadvantages. There may be legal restrictions to limiting the user population. For example, most public transportation systems must accommodate handicaps. Limiting the user population reduces a commercial product's market.

If you do decide to limit your product user, base the limitation on the task analysis described in Sec. 9.3. Determine which tasks have the highest demands on the user and decide the user limits your system will accept. The limits may be skills, age, or physical size and strength. Be certain your product clearly defines these limits.

Training is another way to change how people use your system. Again, industry, the military, and commercial products all use training. Use the same method to identify the need for training that you use to analyze the tasks. The training should bring your system user to a minimum acceptable skill level. The U.S. Navy has excel-. lent publications on developing training programs. Chapter 12 include information on these publications.

Labels and instruction manuals can change the way people operate and maintain your system. The United States space station plans to provide crewmembers with an enormous computer database of operating instructions. Analyze tasks to determine what the labels and instructions should say. Task with high mental or stress demands usually need instructions or labels to remind the user. The law requires you to put warning labels on items associated with dangerous tasks. The military, industrial associations, and the Department of Labor have guidelines to help you format labels and instruction manuals. See Chapter 12.

Labels may be less expensive than design changes, but this is not always the best solution. One example is a matchbook. Matchbooks used to have the warning to "Close cover before striking." The striking surface was on the front of the matchbook so that the user had to close the cover to prevent a spark from igniting all the matches in the book. This could cause serious burns. However, people ignored the warning because it became so familiar. Eventually some genius put the striking surface on the opposite side from the cover. Now, even when the cover is open, there is less chance of igniting all the matches. The simple design solution made the matchbook inherently safer and left more advertising room on the cover.

Human factors design input should help bridge the gap between the user and the equipment. Does the design assure that the user will be safe and productive? You may not be able to include all the design changes needed to improve the product. However, your human factors analysis will supply you with information that can be turned into improvements in instructions and warning labels. Remember that anything you do to help the user will increase product acceptance and therefore increase your design's image.

10

Where Do I Find Human Factors Information?

10.1 The Process for Finding Human Factors Information

Anyone responsible for designing equipment people use and maintain will need information about human size, shape, and capabilities. Figure 10.1 diagrams a process for finding human factors information. Each step in the diagram refers to topics in this book. Use the diagram to help you find the information you need. We will briefly introduce the diagram in this section. The sections that follow discuss each step in more detail.

In Fig. 10.1 the first decision you must make is whether you want to hire a human factors professional. Assuming that you do not hire a professional, move on to the next decisions. If you do have design standards for your product locate and read these standards. Government and industry standards have human factors information. Look here first. You may find information you can use while meeting your contractual and legal obligations.

Next, consult human factors design references and handbooks. General human factors design guidelines are becoming more common. Human factors design handbooks are available from the government, professional societies, libraries, and bookstores. Handbooks cover a wide range of topics and may have the information you need.

If you still do not find an answer, you can go directly to information resources. Information resources include research study re-

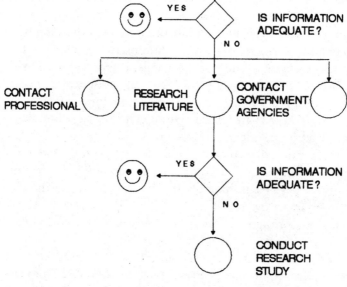

Figure 10.1 Processing for finding human factors information

ports, experts (who are usually flattered to be asked), and government agencies using your tax money to gather and dispense information. Don't be surprised if you get down to the bottom of the Fig. 10.1 and have not found an information source. You can design and conduct simple tests yourself to improve specific design problems. You may feel your study or tests lack the finesse and detail of a research institute's undertaking. However, almost any human factors improvement will make your product more desirable to the consumer or user.

10.2 First Look at Your Design Requirements

Human factors standards contain design information your product should meet. The table in Fig. 10.2 lists standards that may apply to your product and human factors information in the standards. The table also lists which standards are usually required by contract or law and which are voluntary.

STANDARD	NEED FOR COMPLIANCE	TYPE OF HUMAN FACTORS INFORMATION
Government Standards	Mandatory for products designed for the U. S. Government	Detailed and broad human factors data
Occupational Health and Safety Regulations	Mandatory for facilities and equipment used by employees	Data on the required size and location of guards and safety devices on industrial equipment. Environmental limits for worker safety (noise, radiation, illumination, etc.)
Building Codes	Mandatory for structures within specific regions	Minimum sizes for rooms, hallways, and doors. Minimum heating, illumination, and ventilation requirements.
Voluntary Consensus and Trade Standards	Voluntary	Variety of specialized human factors information, including: • Environmental safety criteria • Design standards for specific products (ladders, automobile controls, video display terminals, communications equipment, etc.) • Human factors test methods to measure equipment and the environment

Figure 10.2   Standards are a good source of human factors information.

Use these standards as your first information resource. If you use design information from standards you will also meet your contractual obligations. Manufacturers usually play a big role in developing standards so the information is usually very reliable. If the information is not accurate or not usable, manufacturers will work to remove the faulty information from the standard. Chapter 12 lists specific design standards and how to order them.

10.3 Using Human Factors Design Handbooks

Human factors design handbooks are becoming increasingly popular. They are similar to other engineering handbooks. They supply a variety of data in graphic and tabular format. Recently, many human factors design handbooks are concentrating entirely on individual products like computer hardware and software.

Human factors design handbooks are available in libraries and large bookstores. In college and university bookstores, human factors books can be found with the textbooks for psychology and engineering courses. Even if your design does not have to meet them, U. S. government standards are very useful. Human factors standards published by the Department of Defense and NASA are so comprehensive that they could be considered design handbooks. They are also inexpensive.

10.4 On-Line Literature Searches

Many human factors studies never get into handbooks and standards. These studies usually focus on specific problems and they may have the exact information you need. Besides the manual literature searches, human factors information can be obtained from computer searches. Physicians, attorneys, and other disciplines use electronic research to supply their information needs.

The diagram in Fig. 10.3 shows how you receive information from databases. First a database provider (a private, public, or government organization) collects information from various sources. These sources include journal articles, conferences, books, reports, or government publications. The database provider sends this information to an information service in a machine-readable format. Next an information service loads all these databases into computers. A

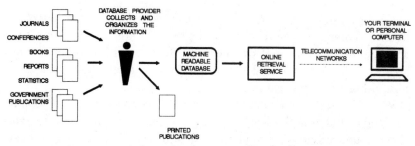

Figure 10.3 How you get information from an on-line database.

nationwide or worldwide telecommunications network connects your computer to an information service. Using a modem connected to your computer, you call the telecommunications network which connects you to the information service. You then decide which databases you wish to search and conduct your search.

There are several ways for you to obtain the information you need from the databases: libraries, information brokers, or your own search. Libraries often will complete a basic search free, however, they often are not as complete as you will need for your design improvement. Your success depends on two things: (1) Does the library have access to the databases you need? (2) Do they have the skills or budget to supply a complex search?

Another choice is to hire an information broker to do your search. Information brokerage companies are in most cities. You can find these companies listed in the phone book under headings like "Library Research Services" or "Information Services." If you have difficulty in locating an information broker your company, university, or public library may know companies doing on-line research in your area.

Information brokerage companies charge for planning time (establishing a search), actual on-line time, database time ($60 per hour and up), and the cost of any publications ordered. Although database time can be high per hour, a simple well-planned search can take as little as 5 to 10 min on-line.

The final choice is to do the search yourself. If you wish to do your own search there are books listing available databases and services. If you have a military contract it may be possible to use the Defense Technical Information Center (DTIC). They conduct searches for you or, with training, allow you to do your own searching.

Many information services conduct training classes or can supply training manuals. Services such as Compuserve or Knowledge Index have menu-driven formats to lead you though a search on-line. The menu-driven format can be very slow and difficult for a complex search. With menu-driven formats you do not have to read a manual or attend a training course.

We use Dialog Information Services to conduct most of our searches. Dialog is the world's largest on-line information retrieval service and provides a broad scope of subjects. You can search over 380 different databases with this service.

The following is an example database search. The first step in a search is to determine which databases you are going to search. Human factors is such a large discipline and there are unlimited databases containing information about humans and the designs humans use. It is important to know which databases contain information you can use to improve your product design. The on-line services tell you what subjects are in each database. After you have determined which databases you want to use, you must plan a search. This is important even if you have hired an information broker to conduct your search.

Figure 10.4 shows how to plan a search on Dialog. The first part of the plan is to decide what the concepts will be. For example, you want to learn about artificial gravity and the human factors effects. You have three concepts: artificial gravity, human factors, and ef-

CONCEPT A (AND)	CONCEPT B (AND)	CONCEPT C
artificial gravity gravitational effects Variable gravity reduced gravity Coriolis effects centrifugal force centripetal force	human factor human factors manned system manned systems man machine system man machine systems human engineering ergonomics man machine engineering	effect effects human performance space flight physiological effects psychomotor performance health

OR

Figure 10.4   How to plan a search on Dialog.

fects. You can have as many concepts as you wish. You start by deciding different terms for artificial gravity and list them under concept A. Next you determine which terms you wish to use for human factors and list them under concept B. Finally you decide what other words to use for the word "effects" and list them under concept C.

Now you enter your search into your computer and your computer "talks" to Dialog's computer. Your computer requests Dialog's computer to match your search with over 130 million records, or information items. Often these records (depending on the specific database you are in) are a bibliographic reference. You are able to review the results of your search on-line or have a printout sent to you. If you find a report you want you can order it while on-line.

The synonyms for human factors under concept B will help you or your researcher conduct other human factors searches. You must use correct terms to obtain the best results from each database. Some databases like the military databases will use the term "human engineering," while computer equipment databases often use the term "ergonomics."

10.5   Contact Professionals

You can obtain human factors information from human factors, medical, and psychological professionals. You can find human factors professionals by contacting the following:

Universities or colleges

Industry (particularly industries with military contracts)

Professional societies

Chapter 12 list colleges and universities and professional societies associated with human factors. Professionals are often able to answer questions specific to your project, or refer you to other people or studies. Most human factors professionals are delighted when someone asks a question about human factors.

10.6   Ask the Government

Some government agencies have human factors information or can direct you to experts. A listing of these agencies is in Chapter 12.

10.7   Do it Yourself

If you reach the final circle on Fig. 10.1 and still do not have the information you need, don't feel alone. Professionals who apply human factors to designs are continually designing and conducting surveys and tests to find information required for a design.

If you have been unsuccessful in finding human factors information you may be tempted to just guess. Do not do this. A short test or survey may provide you with exactly the information you need to improve your design. The table in Fig. 10.5 has helpful advice for your human factors study.

HUMAN FACTORS TEST ELEMENTS	ADVICE
Subjects	Subjects should represent the user of your product or system Test the full range of users: Test a statistically significant number of subjects or, Test subjects that represent the population extremes (as defined by other studies) Use tests to select subjects. Do not make assumptions such as: all large people are strong or small people weak. Use yourself as a subject when planning the test
Task Simulation	Simulate environments that will degrade human performance (cold, hot, dark, etc..) Simulate hardware or software at a level appropriate to answer your question
Scheduling	Try to schedule your test with other tests to save money Schedule well before the time you need the data and plan for rescheduling
Data Collection	Standardize procedures for all trials and subjects Collect data that will answer your question. Collect other data as time and money permit. Don't be distracted. Do not bias results with your presence Record as much as possible for future reference (photos, audio recording, video recording, etc.) Administer the test yourself if possible Wear appropriate clothing Understand the operation of all test equipment Carry spare parts and consumables (batteries, tape, microphones, bulbs, pens)
Data Analysis	Statistical analyses should not surpass your test design (statistics will not improve a poorly designed test) Put results in graphic or tabular format for easy application Analyze and summarize data as soon as possible (before subjects, test equipment, and memory are lost)

Figure 10.5 Helpful advice for a human factors study.

Part

III

References

11

Glossary of Terms

Accelerometer An instrument for measuring vibrations or accelerations.

Activity Center An area where people are "off duty" and not required to do any specific tasks. Examples of activity centers are crew quarters, recreation facilities, galley, and meeting facilities. Workstations are unlike activity centers because people have responsibilities and performance measures in a workstation. Human factors design principles apply to both activity centers and workstations.

Allocate or Allocation of Functions Deciding how functions will be done. Selecting the people, the machines, and the software to do the function. Allocation of functions occurs early in a system design. Human factors should be involved in allocation of functions to assure that humans are efficiently used and not overburdened with responsibilities.

Anthropometry The study of human body measurements.

Artificial Gravity Force simulating gravity force for humans in space. Artificial gravity is the centrifugal force generated when rotating a room around an axis. Artificial gravity is thought to be necessary to preserve the physical health of humans on long-term space missions.

Biomechanical Models Theoretical models of the human body to predict the mechanical behavior.

Biomechanics Science concerning the structure and mechanical behavior of the human body bones and muscles. Biomechanics combines the sciences of anthropometry, mechanics, physiology, and engineering.

Body Envelope The volume envelope which just encloses the body and body motions during an activity.

Comfort That condition of mind which expresses satisfaction with the environment. Human comfort is an important human factors goal. However, comfort is more difficult to measure than other human factors goals such as safety and productivity.

Component   Part of a system. Components can be people, procedures, equipment, software items, or environment.

Consultant   An individual called on for professional advice or opinions. Some consultants specialize in human factors information.

Crew Station   Any location where a crewmember does a task or activity. There are two basic types of crew stations: workstation and activity center.

Deliverable   A tangible item, such as a report or hardware, that a contract requires you to provide your customer.

Design Guidelines   Information to help design a product or system. Design guidelines are never obligatory; unlike design requirements.

Design Handbooks   Collection of information to help design a product or system. The rules and guidelines are usually in tables, charts, or other graphic form to help interpretation and application.

Design Requirements   Rules specifying the form and function of a design. Design requirements are obligatory; unlike design considerations.

Design Standards   Set of guidelines and requirements that the designer should follow when designing a system. Design standards are specific for a system.

Display   Hardware item used to present system information needed by an operator to make decisions for controlling the system.

Dry-Bulb Temperature   Air temperature measured by a common thermometer.

Environmental Control   Control of ambient conditions to produce habitable environments.

Ergonomics   A term used interchangeably with human factors. This term is used worldwide.

Extravehicular Activity (EVA)   Activities performed by a space-suited crewmember in an unpressurized or space environment.

Exposure Limit   Maximum safe environmental exposure for humans. The exposure limit is usually defined by time and a value for an environmental quality. There are exposure limits for noise, radiation, toxic substances, heat, cold, and vibration.

Function   A general purpose or intent to meet a defined mission.

Function Allocation   See Allocate or Allocation of Functions.

Functional Analysis   Breaking the function into smaller parts (subfunctions) and defining the input and output for each subfunction. System designers must do functional analysis before they can allocate the functions to people and equipment.

Human Engineer   A professional title indicating that the individual works in human factors and holds a degree in engineering.

Human Engineering   A term used interchangeably with human factors. This term is commonly used by the United States military.

Human Factors   The process of designing for human use. This term is principally used in the United States.

Human Performance Measurements   Quantifications of human mental or physical performance in different conditions. Human factors researchers measure human performance. Designers and professionals applying human factors measure human performance to evaluate designs.

Hygrometer   An instrument to measure the humidity of the air.

Illumination   The amount of light falling on a surface.

Luminance   The amount of light coming from a source; a measure of brightness.

Manikin   A tool representing human shape, size, and movement range. Manikins can be two or three dimensions, although two-dimension (flat) manikins are more common. Manikins are usually scaled to match engineering drawings. Designers and human factors professionals use manikins. Draftspersons use two-dimensional manikins as templates. Two-dimensional manikins are useful in early design stages for rough approximations.

Man-Machine Engineering   A term used interchangeably with human factors. Used frequently by the United States military.

Microgravity   Condition where gravity effects are small enough to be negligible. People and objects float.

Mockup   A full size, three-dimensional representation of a workstation or activity center. Designers and human factors professionals use mockups to analyze human tasks and evaluate a design before it is completed.

Model   A reduced-scale representation of a hardware item. Models are less useful for human factors studies than mockups. A model is also a three-dimensional representation of the human body in a computer-aided design system. An analyst can use the computer body model to evaluate the size of a workstation and the placement of controls and displays.

Oculometer   An instrument used to track the path of the eye across a surface.

Operation   An organized procedure or process. A sequence of operations comprise a function; operations are subfunctions. An operation can be done by a human, machine, or human-machine combination.

Operational Sequence Diagram (OSD)   A task analysis technique that shows task time, task sequence, and coordination between people and machines.

Percentile   A number indicating the percentage of persons within a population who have a body dimension of a certain size or smaller. For example, the 5th percentile height of males in the U.S. Army is roughly 64 in. This means that 5 percent of the males are 64 in or shorter.

Product   Hardware and software items. This book uses the word "product" to designate individual hardware and software items as opposed to large systems that require coordination of many people and hardware and software items.

Prototype Software   Computer software to simulate the user interface in a full computer program. Prototype software is very useful for human factors analysis in the early stages of software development.

Psychrometer   a tool that measures relative humidity by wet- or dry-bulb method to evaluate human comfort.

Rapid Prototyping   Using computer software to simulate the user interface in a full computer program.

Recommendations   Information contained in a design standard that should be considered but is not necessary to follow exactly.

Requirements   Design standards that must be followed. Research and experience has confirmed that human factors design requirements are necessary for systems to be safe and productive.

Seat Reference Point (SRP)   The point at which the centerline of the seat back surface (depressed) and seat bottom surface (depressed) intersect. The SRP is used to locate controls and displays in a workstation.

Statement of Work   A portion in a contract stating the tasks to be done, the deliverables, and the due dates. Government contracts usually specify human factors requirements in the statement of work.

System   A combination of people, equipment, software, and procedures to accomplish a mission. A system also includes the environment it operates in.

Task   An activity accomplished by a human. Machines do not do "tasks."

Task Analysis   Process to describe tasks in terms of stimulus, equipment used in the task, human response, feedback, and task characteristics such as time and performance requirements. Task analysis is a way to evaluate human performance and compare alternate ways for doing the task. There are many techniques to do task analysis. Task analysis is an essential for human factors to help determine the best combination of people, equipment, procedures, and environment.

Time and Motion Analysis A task analysis technique that focuses on times to do small subtasks. Time and motion analysis is useful in designing for repetitive operations.

Time Line A task analysis technique that focuses on time to do each task step. Time line analysis is useful in designing for operations that must be completed as quickly as possible.

Variable Gravity Another name for artificial gravity. Variable gravity is the force simulating gravity for humans in space. Variable gravity is the centrifugal force generated when rotating a room around an axis. Changing the rotation speed and radius changes the variable gravity force.

Visual Display Terminal (VDT) An electronic device that presents computer-generated visual information. Examples include: cathode ray tube (CRT), liquid crystal diode (LCD), light emitting diode (LED), plasma, and electroluminescent (EL).

Workstation Area where people are required to do specific tasks. Examples of workstations are a desk, automobile driver's seat, airplane cockpit, missile launch facility. Activity centers are unlike workstations because people do not have responsibilities and performance measures in activity centers. Human factors design principles apply to both activity centers and workstations.

Zero Gravity Condition where gravity effects are small enough to be negligible. People and objects float.

Chapter

12

References

Books

Badre, A., and B. Schneiderman (eds.), *Directions in Human/Computer Interaction*, Ablex Publishing, Norwood, N.J., 1982.

Boff, K. R., and Lincoln, J. E. (eds.), *Engineering Data Compendium, Human Perception and Performance*, Harry G. Armstrong Aerospace Medical Research Laboratory Wright-Patterson Air Force Base, Ohio, 1988.

Chaffin, D. B., and G. Anderson, *Occupational Biomechanics*, Wiley, New York, 1984.

Committee on Industrial Ventilation, *Industrial Ventilation, Manual of Recommended Practice*, American Conference of Governmental Industrial Hygienists, P.O. Box 453, Lansing, Michigan.

Damon, A., H. W. Stoudt, and R. A. McFarland, *The Human Body in Equipment Design*, Harvard University Press, Cambridge, Mass., 1966. (Library of Congress Catalog Card No. 65-22067.)

DHHS (NIOSH) Publication No. 81-122, *Work Practices Guide for Manual Lifting*, Department of Health and Human Services, National Institute for Occupational Safety and Health, U.S. Government Printing Office, Washington, D.C., 1981.

Dreyfuss, H., *Symbol Source Book*, McGraw-Hill, New York, 1972. (Library of Congress Card No. 71-172261.)

Eastman Kodak Co., *Ergonomics Design for People at Work—Volume I: Workplace, Equipment, and Environmental Design, and Information Transfer*, Van Nostrand Reinhold, New York, N.Y., 1986.

Eastman Kodak Co., *Ergonomics Design for People at Work—Volume II: Job Design and Manual Handling*, Van Nostrand Reinhold, New York, N.Y., 1986.

Foley, J. D., and A. Van Dam, *Fundamentals of Interactive Computer Graphics*, Addison-Wesley, Reading, Mass., 1982.

Garett, J. W., and K. W. Kennedy, *A Collation of Anthropometry*, (2 Vols.), Aerospace Medical Research Laboratory, Wright-Patterson Air Force

Base, Ohio, 1971. (AD 723 629; Library of Congress Catalog Card No. 74-607818.)

Grandjean, E., *Fitting the Task to the Man: An Ergonomic Approach*, 3d ed., Taylor Francis, London, 1980.

NcNaught, A. B., and R. Callander, *Illustrated Physiology*, Williams & Wilkins, Baltimore, Md., 1963.

Meister, D., and D. Sullivan, *Guide to Human Engineering Design for Visual Displays*, The Bunker-Ramo Corp., Contract No. NOOO1468-C-027E, Work Unit No. NR196-080 (AD 693 237), Office of Naval Research, 1969.

NASA, *Anthropometric Source Book—Volume I: Anthropometric for Designers; Volume II: A Handbook of Anthropometric Data; Volume III: Annotated Bibliography of Anthropometric*, NASA Scientific and Technical Information Office, Yellow Springs, Ohio, 1978

NAVEDTRA 110A, *Procedures for Instructional System Development*, Report 0502-LP-000-5510, Department of the Navy, Chief of Naval Education and Training, Pensacola, Fla., 1981.

Parker, J. F., Jr., and V. R. West, (eds.), *Bioastronautics Data Book*, 2d ed., U.S. Government Printing Office, NASA SP-3006, Washington, D.C., 1973.

Salvendy, Gavriel, (ed.), *Handbook of Human Factors*, Wiley, New York, Dept. 063, 1987.

Schneiderman, B., *Software Psychology: Human Factors in Computer and Information Systems*. Winthrop Publishers, Cambridge, Mass., 1980.

Van Cott, H. P., and R. G. Kinkade (eds.), *Human Engineering Guide to Equipment Design*, Wiley, New York, 1972. (Library of Congress Catalog Card No. 72600054.)

Woodson, W. E., *Human Factors Design Handbook: Information and Guidelines for the Design of Systems, Facilities, Equipment, and Products for Human Use*, McGraw-Hill, New York, 1981.

Organizations and Government Agencies

Air Force Human Resources Laboratory, Library
Brooks Air Force Base, TX 78235
(512)536-2651

Publishes technical reports. Literature collection accessible for on-site use.

American Institute of Aeronautics and Astronautics (AIAA)
557 West 57th Street
New York, NY 10019
(212)247-6500

A professional organization devoted to science and engineering in aviation, space technology, and systems. They publish books, journals, and technical papers.

American Institute of Industrial Engineers (AIIE)
25 Technology Park
Norcross, GA 30092
(404)449-0460

A professional organization that conducts conferences, publishes books, and produces journals and technical papers relating to human factors.

American National Standards Institute (ANSI)
1430 Broadway
New York, NY 10018
(212)354-3300

A corporation composed of organizations, companies from private industry, and government professionals. They answer questions and make referrals to experts. It publishes reports and a directory is available.

American Society of Safety Engineers
1800 E. Oakton Street
Des Plaines, IL 60018-2187
(312)692-4121

A professional society that can provide information on accident prevention. Refers inquires to safety engineers.

Center for Ergonomics, University of Michigan,
1205 Beal Street
Ann Arbor, MI 48109
(313)763-2243

Research and training center that holds training sessions and monthly seminars on occupational ergonomics. The center will make referrals and publishes ergonomics information.

Consumer Product Safety Commission
Human Factors Division
5401 Westbard Avenue
Bethesda, MD 20207
(301)492-6468

The agency will provide evaluation reports and answer questions about human factors and products.

Crew System Ergonomics Information Analysis Center (CSERIAC)
AAMRL/HE/CSERIAC
Wright Patterson AFB
Dayton, OH 45433-6573
(513)255-4842

This information analysis center (IAC) provides human factors information contained in Department of Defense documents. CSERIAC will provide approved users information on human factors.

Defense Technical Information Center (DTIC)
U.S. Department of Defense
Building 5, Cameron Station
Alexandria, VA 22304
(202)274-6871

DTIC is a databank of multidisciplinary information available to U.S. government agencies, contractors, subcontractors, grantees, and universities. DTIC has an interactive database (DROLS). If you cannot qualify to use DTIC a large percentage of DTIC, is available from NTIS (file 6) on Dialog.

Department of Transportation
Transportation System Center
Kendall Square
Cambridge, MA 02142
(617)494-2486

A research department of DOT. Technical reports can be searched on TRIS or be obtained from NTIS.

Dialog Information Services
3460 Hillview Avenue
Palo Alto, CA 94304
(800-334-2564)

Dialog is the largest online information retrieval service with over 380 different databases available to searchers.

Environmental Protection Agency (EPA) Library
Toxic Substances Library
401 M Street, NW
Washington, DC 20460
(202)382-5922

Reference materials, including books, journals, EPA documents, and legal information sources in the field of environmental protection. The materials are available though NTIS. NTIS file 6 is available on Dialog.

European Space Agency (ESA)
955 L'Enfant Plaza, SW, Suite 1404
Washington, DC 20024
(202)488-4158

Eleven European governments combined space program organization (similar to NASA). ESA prepares special reports, conference proceedings, bulletins, a journal, and other documents. This office will answer questions from the public and send copies of publications.

Federal Aviation Administration
Office of Aviation Medicine
AAM-1
800 Independence Avenue SW
Washington, DC 20591
(202)426-3535

Data on civil aviation medicine in aviation safety, occupational health, and biomedical research. Answers questions and supplies publications.

Federal Aviation Administration Technical Center
ACT 624
Atlantic City Airport
Atlantic City, NJ 08405
(609)484-5124

Investigates aviation communication, guidance, safety and air traffic control systems. Answers inquiries.

Federal Transportation Safety Recommendations Data Base
National Transportation Safety Board
800 Independence Avenue, SW
Washington, DC 20594
(202)382-6817

An interactive on-line database containing safety recommendations. Search and printouts are conducted.

Human Factors Society, Inc. (HFS)
P.O. Box 1369
Santa Monica, CA 90406
(213)394-1811

An interdisciplinary professional organization in human factors. The society produces publications and will supply a free publications catalog.

Institute of Electrical Engineers (IEEE)
345 East 47th Street
New York, NY 10017
(212)705-7866 (public information)
(212)705-7890 (technical activities information)

An engineering society concerned with electrical engineering, electronics, computer engineering, and computer sciences.

Institute of Industrial Engineers (IIE)
25 Technology Park Atlanta
Norcross, GA 30092
(404)449-0460

A professional organization providing conferences and publications.

International Organization for Standardization (ISO)
Central Secretariat
1, rue de Varembe
CH-1211 Geneva 20, Switzerland
(011-41-22-34-12-40)

An international agency for standardization. See American National Standards Institute (ANSI) for the U.S. contact.

National Aeronautics and Space Administration (NASA)
NASA Information Center
NAS-2
Washington, DC 20546
(202)453-1000

This office can direct you to the appropriate resource in NASA.

NASA Headquarters Library
Room A-39
600 Independence Avenue, SW
Washington, DC 20547
(202)453-8545

Librarians will help find information and make recommendations.

National Bureau of Standards (NBS)
Laboratories and Centers
Route 270
Gaithersburg, MD 20899
(301)921-1000

NBS provides information on physical measurement standards. Staff members can supply information for obtaining scientific and technological knowledge.

National Institute for Occupational Safety and Health (NIOSH)
Technical Information Branch
4676 Columbia Parkway
Cincinnati, OH 45226
(513)533-8326

An agency that conducts and funds human factors research in safety and occupational health areas. They will supply technical information and answer questions. NIOSH also publishes a bibliographic database (file 160) available on Dialog Information Service.

National Technical Information Service (NTIS)
5285 Port Royal
Springfield, VA 22161
(703)487-4600 or (703)487-4642

The government's central technical information clearinghouse. NTIS provides the public with documents and services obtained from the federal government's research and engineering efforts. NTIS is accessible on-line on Dialog (file 6) and BRS.

National Transportation Safety Board
Office of Government and Public Affairs
800 Independence Avenue, SW
Washington, DC 20594
(202)382-6600

The National Transportation Safety Board has two divisions of professionals who deal with human factors in relation to transportation equipment. Questions should be directed to the above address. Individual accident reports may be purchased from the National Technical Information Service (NTIS).

Nuclear Regulatory Commission (NRC)
Human Factors Safety Division
Washington, DC 20555
(202)492-9530 (Publication Ordering Office)
(301)492-9595 (Division of Human Factors Safety)

This office oversees the following branches below.

(301)492-4816 (Licensee Qualifications Branch)

This office oversees the qualifications and training of nuclear power plant operators.

(301)492-4813 (Human Factors Engineering Branch)

Occupational Safety and Health Administration (OSHA)
U.S. Department of Labor
Room N 3637
200 Constitution Avenue, NW
Washington, DC 20210
(202)523-8151

This agency sets and enforces standards for employees. OSHA maintains a database and will supply pamphlets on topics such as toxicity, safety hazards, and protective equipment.

Special Interest Group on Computers and Human Interaction (SIGCHI)
Association for Computing Machinery
11 West 42d Street
New York, NY 10036
(212)869-7440

SIGCHI studies human-computer interaction processes. Members can answer questions and make referrals.

Society of Automotive Engineers (SAE)
400 Commonwealth Drive
Warrendale, PA 15096
(412)776-4841

A professional society of engineers involved in preparing standards, specifications, and test procedures for ground and space vehicles.

Transportation Research Information Services On-line (TRIS)
Transportation Research Board (TRB)
2100 Constitution Avenue, NW
Washington, DC 20418
(202)334-3250

This database contains ongoing transportation research projects. The database contains bibliographical citations of journals, technical papers, and articles. TRIS-ON-LINE can be accessed in file 63 of Dialog.

U.S. Army LABCOM
Human Engineering Laboratory
Aberdeen Proving Ground, MD 21005-5001
(301)278-5820

Conducts human factors research and provides human factors engineering to acquisition of Army materiel.

Specifications, Standards, and Guidelines

A9.1 Building Exits Code (NFPA 101)
Available at a nominal cost from ANSI
1440 Broadway, New York, NY 10018

A11.1 Practice for Industrial Lighting
Available at a nominal cost from ANSI
1440 Broadway, New York, NY 10018

AFSC DH 1-3 Human Factors Engineering
Request for handbook directed to 4950/TZHM
Wright-Patterson AFB, Dayton, OH 45433

AFSC DH 2-2 Crew Stations and Passenger Accommodations
Request for handbook directed to 4950/TZHM
Wright-Patterson AFB, Dayton, OH 45433

ANSI/HFS Standard No. 100-1988 American National Standard for
  Human Factors Engineering of Visual Display Terminal Workstations
Availble from the Human Factors Society
P.O. Box 1369, Santa Monica, CA 90406

DOD-HDBK-763 Human Engineering Procedures Guide
Available from the Standardization Documents Order Desk, Building 4D
700 Robbins Avenue, Philadelphia, PA 19111-5094

ESD-TR-86-278 Guidelines for Designing User Interface Software
Available from the Naval Publications and Forms Center (ATTN:
NPODS)
5801 Tabor Avenue, Philadelphia, PA 19120-5099

Human Engineering Guide to Equipment Design
Available from Superintendent of Documents, U.S. Government Printing
Office, Washington, DC 20402

ISO DIS 2631 Guide to the Evaluation of Human Exposure to Whole
Body Vibration
Available from American National Standards Institute, Inc.
1430 Broadway, New York, NY 10018

Lighting Handbook
Available from the Illumination Engineering Society
345 East 47th Street, New York, NY 10017
(212)705-7926

MIL-H-1472D Human Engineering Design Criteria for Military Systems,
Equipment and Facilities
Available from the Standardization Documents Order Desk, Building 4D
700 Robbins Avenue, Philadelphia, PA 19111-5094

MIL-H-46855B Human Engineering Requirements for Military Systems,
Equipment and Facilities
Available from the Standardization Documents Order Desk, Building 4D
700 Robbins Avenue, Philadelphia, PA 19111-5094

MIL-HDBK-759A Human Factors Engineering Design for Army Materiel
Available from the Standardization Documents Order Desk, Building 4D
700 Robbins Avenue, Philadelphia, PA 19111-5094

MIL-HDBK-761A Human Engineering Guide for Management
Information Systems
Available from the Standardization Documents Order Desk, Building 4D
700 Robbins Avenue, Philadelphia, PA 19111-5094

MIL-STD-12 Abbreviation for Use on Drawings, Specifications,
Standards, and in Technical Documents
Available from the Standardization Documents Order Desk, Building 4D
700 Robbins Avenue, Philadelphia, PA 19111-5094

MIL-STD-882 System Safety Program Requirements
Available from the Standardization Documents Order Desk, Building 4D
700 Robbins Avenue, Philadelphia, PA 19111-5094

NASA-STD-3000, Man-System Integration Standards
Request from MSIS Custodian/SP34
NASA-Johnson Space Center, Houston, TX 77058

NASA-SP-3006 Bioastronautics Data Book, 2d ed., J. F. Parker and V. R.
West, (eds.)
Available from Superintendent of Documents, U.S. Government Printing
Office, Washington, DC 20402

SAE J925 Minimum Access Dimension for Construction and Industrial
Machinery
Available from Society of Automotive Engineers
400 Commonwealth Drive, Warrendale, PA 15096-001

Standard 55-81 Thermal Environment Conditions for Human Occupancy
Available at a nominal cost from American Society of Heating,
Refrigerating and Air-conditioning Engineers (ASHRAE)

Standard 62-81Ventilation for Acceptable Indoor Air Quality Guide and
Data Book
Available at a nominal cost from American Society of Heating,
Refrigerating and Air-conditioning Engineers (ASHRAE) 1791 Tullie
Avenue, NE
Atlanta, GA 30329

Threshold Limits Values
Available from American Conference of Governmental Industrial
Hygienists (ACGIH)
1014 Broadway, Cincinnati, OH 45202

29 CFR 1910 Occupational Safety and Health Standards (Federal
Regulation)
Available from Superintendent of Documents, U.S. Government Printing
Office
Washington, DC 20402

Conferences

Human Factors Society
October 1990—34th Annual Meeting in Orlando, Florida

International Ergonomics Association
July 1991—11th IEA Congress in Paris
August 1994—12th IEA Congress in Toronto

Schools and Universities

This is a list of schools and universities offering courses or degrees in human factors. These courses and programs are continually changing and we have not evaluated these programs. Many of these schools have research programs and authorities that can help you in your design.

California State University, Northridge
Degrees: M.A. (Psychology)
Address: Human Factors Program, Department of Psychology, CSUN,
18111 Nordhoff Street, Northridge, CA 91330
(818)885-2827

George Mason University
Degrees: B.S. (Industrial Psychology); M.A. (Industrial Psychology); Psy.
D. (Applied Psychology—Human Factors)
Address: Psychology Dept., George Mason University
4400 University Drive, Fairfax, Virginia 22030
(703)323-2207/2203

Massachusetts Institute of Technology
Degrees: M.S.; Ph.D. (Man-Machine Systems)
Address: Department of Mechanical Engineering, Massachusetts Institute
of Technology
Cambridge, MA 02139

New Mexico State University
Degrees: M.A. (Engineering Psychology), Ph.D. (Engineering Psychology
or applied cognitive)
Address: Department of Psychology, New Mexico State University
Box 3452, Las Cruces, NM 88003
(505)646-2502

Purdue University
Degrees: B.S., M.S., Ph.D. (Industrial Engineering)
Address: School of Industrial Engineering
W. Lafayette, Indiana 47907
(317)494-5426

Degrees: B.S., M.S., Ph.D. (Psychology)
Address: Department of Psychological Sciences, Purdue University
W. Lafayette, Indiana 47907
(317)494-5426

Texas A&M University
Degrees: M.S. (Engineering); Ph.D. (Engineering)
Address: Industrial Engineering, Texas A&M University
College Station, Texas 77843
(409)845-5403, 846-0837

University of Michigan
Degrees: M.S.E., M.S., Ph.D.; Industrial and Operations Engineering
Address: Center for Ergonomics, The University of Michigan
1205 Beal Street
Ann Arbor, Michigan 48109
(313)763-2245

University of Southern California
Degrees: MSOSH; MSOH; MSSM
Address: Human Factors Department, University of Southern California
Los Angeles, California 90089-0021
(213)743-7915

University of Wisconsin-Madison
Degrees: M.S. Industrial Engineering; Ph.D. (Industrial Engineering)
Address: Department of Industrial Engineering
Room 309, 1513 University Avenue, Room 390, Madison, Wisconsin
53706
(608)263-6329

Virginia Polytechnic Institute and State University
Degrees: M.S. and Ph.D. in IEOR with a concentration in human factors.
Address: Virginia Polytechnic
302 Whittemore Hall, Blacksburg, Virginia 24061
(703)961-6656

Wright State University
Degrees: M.A. (Applied Behavioral)
Address: Applied Behavioral Science, Wright State University
Dayton, OH 45435
(513)873-2391

Degrees: B.S. (Human Factors Engineering)
Address: Human Factors Engineering, Wright State University
Dayton, OH 45435
(513)873-2701

Journals

Aerospace America, (monthly), American Institute of Aeronautics and Astronautics (AIAA), 557 West 57th Street, New York, NY 10019

American Industrial Hygiene Association Journal, (monthly), American Industrial Hygiene Association, 475 Wolf Ledges Parkway, Akron, OH 44311

Applied Ergonomics, (quarterly), IPC House, Surrey, England

British Journal of Industrial Medicine, (quarterly), British Medical Association House, Tavestock Square, London WCIH 9JR, England

British Medical Journal, 1172 Commonwealth, Boston, MA 02134.

Ergonomics, (bimonthly), Taylor Francis, Ltd., 10-14 Macklin Street, London WC2B, 5NF, England

Ergonomics Abstracts, (quarterly), Taylor Francis, Ltd., 10-14 Macklin Street, London WC2B, 5NF, England

European Journal of Applied Physiology, (quarterly), Springer-Verlag, 165 Fifth Avenue, New York, NY 10010

Human Factors Journal, (bimonthly), Human Factors Society, Santa Monica, CA 90406

Journal of Occupational Medicine, (monthly), American Occupational Medical Association, Arlington Heights, IL 60005

SIGCHI Bulletin, (quarterly), Special Interest Group on Computers and Human Interaction, 11 West 42d Street, New York, NY 10036

Index

ABOUT THE AUTHORS

PEGGY and BARRY TILLMAN are the principals of Tillman Ergonomics Company. They have extensive experience working with private organizations and government agencies.

PEGGY TILLMAN has degrees in Experimental Psychology and Teaching. Her specialties include research, training, and consumer ergonomics. She is a member of the American Institute of Aeronautics and Astronautics and the Human Factors Association of Canada. Peggy Tillman has been President and Manager of Tillman Ergonomics since 1983.

BARRY TILLMAN has worked as a Human Factors Engineer since 1967 and has managed many large human factors programs. He has a master's degree in Industrial and Systems Engineering. He is a member of the American Institute of Aeronautics and Astronautics and the Human Factors Society.

PEGGY and BARRY TILLMAN are authors of numerous professional papers and standards. They are coauthors of *Human Factors Design Handbook* (second edition, 1991), published by McGraw-Hill.

The Tillmans can be contacted at Tillman Ergonomics Company, Inc., P. O. Box 165, Fox Island, WA 98333.